Springer Climate

Series Editor
John Dodson, Institute of Earth Environment, Chinese Academy of Sciences, Xian, Shaanxi, China

Springer Climate is an interdisciplinary book series dedicated to climate research. This includes climatology, climate change impacts, climate change management, climate change policy, regional climate studies, climate monitoring and modeling, palaeoclimatology etc. The series publishes high quality research for scientists, researchers, students and policy makers. An author/editor questionnaire, instructions for authors and a book proposal form can be obtained from the Publishing Editor.

Now indexed in Scopus® !

More information about this series at https://link.springer.com/bookseries/11741

Claus Kondrup · Paola Mercogliano ·
Francesco Bosello · Jaroslav Mysiak ·
Enrico Scoccimarro · Angela Rizzo · Rhian Ebrey ·
Marleen de Ruiter · Ad Jeuken · Paul Watkiss
Editors

Climate Adaptation Modelling

Springer

Editors
See next page

ISSN 2352-0698 ISSN 2352-0701 (electronic)
Springer Climate
ISBN 978-3-030-86213-8 ISBN 978-3-030-86211-4 (eBook)
https://doi.org/10.1007/978-3-030-86211-4

This Springer imprint is published by the registered company Springer Nature Switzerland AG
The registered company address is: Gewerbestrasse 11, 6330 Cham, Switzerland

Editors
Claus Kondrup
DG CLIMA Adaptation Unit
Auderghem, Belgium

Francesco Bosello
Fondazione CMCC—Centro
Euro-Mediterraneo sui Cambiamenti
Climatici
Lecce, Italy

Enrico Scoccimarro
Fondazione CMCC—Centro
Euro-Mediterraneo sui Cambiamenti
Climatici
Lecce, Italy

Rhian Ebrey
Department of Water and Climate Risk
Institute for Environmental Studies
Amsterdam, The Netherlands

Ad Jeuken
Deltares
Delft, The Netherlands

Paola Mercogliano
Fondazione CMCC—Centro
Euro-Mediterraneo sui Cambiamenti
Climatici
Lecce, Italy

Jaroslav Mysiak
Fondazione CMCC—Centro
Euro-Mediterraneo sui Cambiamenti
Climatici
Lecce, Italy

Angela Rizzo
Fondazione CMCC—Centro
Euro-Mediterraneo sui Cambiamenti
Climatici
Lecce, Italy

Marleen de Ruiter
Department of Water and Climate Risk
Institute for Environmental Studies
Amsterdam, The Netherlands

Paul Watkiss
Paul Watkiss Associates
Oxford, UK

Preface

In the last 20 years, the study on climate change adaptation has shown important progress in the knowledge of climate change impacts, inthe methodologies and modelling tools used for adaptation assessment and in the mainstreaming and implementation of adaptation decisions in policy planning.

The vast majority of the monographic literature on adaptation divides into a more technical-sectoral stream, analyzing the main features of adaptation action in specific contexts (e.g. water sector, disaster risk reduction, agriculture, developing countries, etc.) and suggesting guidelines for its implementation, and a more theoretical-conceptual one, describing the multidimensional challenges for adaptation assessment and the methodologies to address them.

This book focuses on an issue only marginally tackled by this literature: the still existing gap between adaptation science and modelling and the possibility to effectively access and exploit the information produced by policy making at different levels: international, national and local. To do so, the book presents the proceedings of a high-level expert workshop on adaptation modelling, integrated with main results from the "Study on Adaptation Modelling" (SAM-PS) commissioned by the European Commission's Directorate-General for Climate Action (DG-CLIMA) and implemented by the CMCC Foundation—Euro-Mediterranean Centre on Climate Change, in collaboration with the Institute for Environmental Studies (IVM), Deltares, and Paul Watkiss Associates (PWA).

More specifically, the book presents: results of an up-to-date overview of the technical, financial, economic and non-monetary models and tools for climate change hazards, exposure, vulnerability, and adaptation assessments, highlighting main research gaps with emphasis on the interaction with and usability by the policy community. Building on selected use-cases and representative case studies, it also proposes practical approaches to conduct adaptation assessments as guidelines for decision makers at different levels of governance. All issues are enriched by the contributions of the high-level expert workshop, where authoritative experts in the field of adaptation, from the academia and public agencies, discuss the interconnections between the science and policy of adaptation, identify challenges and latest development from adaptation modelling and offer suggestions to overcome the gaps.

This book, mainly addressed to academics, policy makers and practitioners in the field of adaptation, aims to: providing orientation in the large and expanding methodological/quantitative literature, presenting novelties, guiding in the practical application of adaptation assessments and suggesting lines for future research.

Auderghem, Belgium — Claus Kondrup
Lecce, Italy — Paola Mercogliano
Lecce, Italy — Francesco Bosello
Lecce, Italy — Jaroslav Mysiak
Lecce, Italy — Enrico Scoccimarro
Lecce, Italy — Angela Rizzo
Amsterdam, The Netherlands — Rhian Ebrey
Amsterdam, The Netherlands — Marleen de Ruiter
Delft, The Netherlands — Ad Jeuken
Oxford, UK — Paul Watkiss

Introduction from DG-CLIMA

The European Commission adopted the **new EU Strategy**[1] *on adaptation to climate change* on 24 February 2021, setting out the pathway to prepare for the unavoidable impacts of climate change. While the EU does everything within its power to mitigate climate change, domestically and internationally, we must also get ready to face its unavoidable consequences. The **climate crisis** already affects every place in Europe. Climate change is happening today, so we have to build a more resilient tomorrow. The world has just concluded the hottest decade on record during which the title for the hottest year was beaten eight times. The frequency and severity of climate and weather extremes is increasing. These extremes range from unprecedented forest fires and heatwaves above the Arctic Circle to devastating droughts in the Mediterranean region, and from hurricanes ravaging EU outermost regions to forests decimated by unprecedented bark beetle outbreaks in Central and Eastern Europe. Slow onset events, such as desertification, loss of biodiversity, land and ecosystem degradation, ocean acidification or sea level rise are similarly destructive over the long term. Building on the 2013 Climate Change Adaptation Strategy, **the aim of the new strategy** is to shift the focus from understanding the problem to developing solutions and to move from planning to implementation. To inform adaptation action and decision-making, **smarter adaptation** requires more data collection and data sharing on a wider range of topics including climate-related risks and losses, and health.[2] Adaptation actions must be informed by robust data and risk assessment tools that are available to all—from families buying, building and renovating homes to businesses in coastal regions or farmers planning their crops. To achieve this, the strategy proposes actions that push the frontiers of knowledge on adaptation. Climate change has impacts at all levels of society and across all sectors of the economy. **More systemic adaptation** includes the local level and vulnerable communities, promotion of nature-based solutions, and topics such as fiscal policies, for instance, 40% of European GDP is generated in coastal areas exposed to sea level rise. The Commission will continue to incorporate climate resilience considerations

[1] https://ec.europa.eu/clima/policies/adaptation/what_en#tab-0-2.

[2] https://climate-adapt.eea.europa.eu/observatory.

in all relevant policy fields. It will support the further development and implementation of adaptation strategies and plans. **Faster adaptation** includes bridging the gap between planning and implementation, for instance, working to boost adaptation financing, engaging with the insurance sector to reduce the climate protection gap, where the financial burden of natural disasters disproportionally falls on uninsured families and businesses or public finances, and coordinate closely with the new Horizon Europe Mission[3] on adaptation to advance the development, roll-out and scale-up of innovative solutions to make adaptation an easier and more readily available choice for both the public and the private sector. More is needed at the **international level**, where EU climate change adaptation policies must match our global leadership in climate change mitigation. The Paris Agreement established a global goal on adaptation and highlighted adaptation as a key contributor to sustainable development. The EU will step up international action, promote sub-national, national and regional approaches to adaptation, with a specific focus on adaptation in Africa and Small Island Developing States. We will increase support for international climate resilience and preparedness through the provision of resources, by prioritizing action and increasing effectiveness, through the scaling up of international finance and through stronger global engagement and exchanges on adaptation. We will also work with international partners to close the gap in international climate finance. In parallel with the preparation of the new strategy, the Commission undertook an elaborate study on **adaptation modelling**, i.e. the technical, financial, economic and non-monetary analysis and modelling of climate change hazards, risks, impacts, vulnerability and adaptation. The overall objective is to support **better informed decision-making** on adaptation. This book is a key result of the study. The EU Adaptation Strategy emphasises the importance of adaptation modelling, for instance, in the nexus between climate hazards and socioeconomic vulnerability and inequality, and the need to advance on adaptation modelling, risk assessment and management tools, as well as dissemination such as the Climate-ADAPT[4] platform. Climate resilience decision support systems and technical advice must become more accessible and rapid to foster their take-up at all levels of governance and decision-making. The Commission will support the development of rapid response solutions for decision makers and enrich the toolbox for adaptation practitioners. **This book shows the wide scope, relevance and importance of 'adaptation modelling'.**

[3] https://ec.europa.eu/info/horizon-europe/missions-horizon-europe/adaptation-climate-change-including-societal-transformation_en.

[4] https://climate-adapt.eea.europa.eu/.

Introduction from the Editors

The publication of a scientific volume concerning the main results of the "Study on Adaptation Modelling" (SAM-PS) as well as the outcomes of the workshop organized during the last period of the work, was one of the main activities foreseen in the project. Specifically, the study, commissioned by the European Commission's Directorate-General for Climate Action (DG-CLIMA) and implemented by the CMCC Foundation—Euro-Mediterranean Centre on Climate Change, in collaboration with the Institute for Environmental Studies (IVM), Deltares, and Paul Watkiss Associates (PWA), was aimed at:

- conducting a review of climate change adaptation modelling methods and tools (Project's Task 2);
- developing guidance for their application (Project's Task 3);
- providing use-cases as operative examples for their rapid deployment (Project's Task 4);
- organizing a high-level expert workshop (Project's Task 5).

The book describes all the activities carried out during the project in the frame of each project task. In detail, one of the main outcomes of the project is represented by a comprehensive review, which provided an up-to-date and forward-looking overview of the range of technical, financial, economic and non-monetary models and tools for hazards, risks, impacts, vulnerability and adaptation climate assessments. This overview was aimed at addressing the EU Commissions' requirement to support greater-informed decision-making on climate adaptation at multiple governance levels. The overview provided not only the state of the art of the current knowledge on climate adaptation assessment methodologies, but it has also allowed highlighting the main research and policy gaps.

Based on the results of the desk review and accounting for a number of use case and specific case studies, the project team has worked at the proposal of a number of recommended approaches for modelling the climate-related risks and impacts and for analysing the effectiveness of suitable adaptation strategies in order to better inform decision-making on climate adaptation at various levels of governance. These aspects represented the main activities carried out in Task 3 and Task 4.

Furthermore, during the project, a high-level expert workshop was organized to offer the opportunity for modellers and experts from academia, public agencies, business enterprises, and societal organisations to exchange knowledge and good practices on adaptation to climate change and strengthen professional networks. The workshop has also allowed leading the foundations for further transnational cooperation.

The main outcomes of the project, as well as all the contributions presented during the sessions of the workshop and the related highlights, represent the main topic of this volume, which is organized as follows:

Part I This more conceptual and perspective section introduces the major challenges for adaptation modelling. The section opens with an introduction containing a summary of reflections from the study on adaptation modelling presenting a conceptual structure for adaptation assessment and highlighting areas where improvements are needed. A set of expert contributions from the workshop provide further insights on the role of adaptation modelling, the room for their application and applicability, and their final assessment.

Part II This section introduces scope, contributions and the last developments of the modelling literature that addresses the investigation phases which are preliminary, but fundamental to adaptation assessment proper: the investigation of hazard, exposure and vulnerability. The opening introduction presents the findings from the study on adaptation modelling highlighting, in particular, strengths and weaknesses of modelling approaches in this domain and what remains to be done. Then, expert contributions present insightful methodological and modelling examples related to the different preliminary phases of the adaptation assessment ranging from the use of climate data and climate services for adaptation, to hazard and risk assessment in the specific areas of coastal protection, agriculture, flooding, and urban context.

Part III This section is specular to the previous one, but with a specific focus on adaptation modelling and adaptation assessment, considering thus what occurs after the risk assessment phase. As before, the introduction to the section presents the findings from the study on adaptation modelling highlighting, strengths and weaknesses of modlling approaches in this domain, and what remains to be done. The section is then populated by workshop contributions chosen to offer applied and up-to-date examples of adaptation modelling assessment and methods in different key areas: the land system, water systems, infrastructure, coastal areas, health and insurance.

Part IV This section discusses the linkages between adaptation modelling and policy action, focussing, in particular, on those steps and improvements in the former that will facilitate use and uptake from the latter. The introduction presents the findings from the study on adaptation modelling highlighting not only gaps, but also practical suggestions for advancements. A particular emphasis is placed on the need for "rapid analysis" from decision makers and when and how this can be effectively feasible through adaptation modelling. The introduction will also propose some "fiches" where practical demonstration examples of "rapid analyses" are presented to the reader. Workshop contributions support and enrich the initial reflections with specific examples on the relation between adaptation action and the policy dimension. Among the issues tackled there are: the role of uncertainty, the operationalization of climate proofing in policy making, the relation between adaptation and mitigation policies.

Contents

Part I
Challenges for Adaptation Modelling

Paul Watkiss
Paul Watkiss Associates Ltd, Oxford, Oxfordshire, United Kingdom
paul_watkiss@btinternet.com

Introduction

There is now a long and well-established field of adaptation assessment and modelling, stretching back several decades to the early IPCC assessment reports. However, as we enter the decade of the 2020s—and start to plan for climate risks towards the middle of the century—adaptation needs are changing. Adaptation is moving from the theoretical to the applied, and this requires modelling that supports informed decision-making for implementation.

At the same time, our experience of modelling adaptation over the years has provided many lessons and has identified a number of challenges. These include the following:

· Adaptation does not involve a single quantitative objective and it cannot be assessed with a single common metric that applies to all risks and sectors (UNEP 2018). This makes it very different from mitigation.

· Adaptation involves complex temporal dimensions, as it seeks to address impacts that vary dynamically and non-linearly over time. It also has to consider action against a background of high uncertainty (Wilby and Dessai 2010). This uncertainty relates to future emission scenarios, climate model projections, socio-economic scenarios, the physical impacts of climate change, and therefore, the potential benefits of adaptation. These temporal and uncertainty issues make modelling complex.

· Adaptation involves a range of actions, at different aggregation scales (from European level to local). The benefits of this adaptation are often difficult to model, in quantitative terms, because it involves reductions in impacts that are sector-, location- and context-specific. Furthermore, adaptation involves a mix of types of interventions, including non-technical options such as capacity building that are more difficult to model, but which are critical for effective adaptation. The prioritisation of adaptation options requires extended decision support methods (Chambwera et al. 2014), which include consideration of risk and uncertainty.

· Adaptation is governance- and problem-specific (Moser and Ekstrom 2010). There is a multitude of different users who are looking at different adaptation challenges, and who use different frameworks, methods and models. Related to this, adaptation is often implemented through a mainstreaming approach—where it is integrated into policy, plans and projects—rather than as a stand-alone activity (OECD 2015). Adaptation assessment, therefore, needs to consider the wider policy landscape and other factors, not just climate, and address multiple objectives.

· Adaptation is starting to happen in Europe. This provides a new modelling challenge, as this existing adaptation needs to be integrated into modelling baselines, in order to allow the analysis of the benefits of additional action.

· Finally, there are important constraints and barriers to adaptation, which make it difficult for individuals, businesses and governments to plan and implement actions (Klein et al. 2014). These include economic, political economy and governance barriers. These need to be considered alongside modelling assessments.

In the past, adaptation studies have somewhat avoided these challenges by using stylised technical modelling to assess options. Such studies are extremely valuable in raising awareness and providing early policy-relevant information, but they do not provide the information for real-world adaptation, especially for decision makers who need to act now (not in 2050).

Very positively, the area of adaptation modelling is evolving rapidly. Modelling studies are recognising and seeking to overcome these challenges. The papers in this part discuss the above mentioned challenges and identify or propose solutions to many of them.

The part starts with a paper by Schwarze et al. that focuses on one particular area of climate modelling, the global integrated assessment models, and reviews how they consider adaptation. The paper recommends a new phase of "aging and learning" of adaptation models to better fit the heterogeneity of adaptation measures. Aaheim et al. look at the cross-sectoral challenges for adaptation modelling and look at another set of global models, the multi-region, multi-sector computable general equilibrium (CGE) models, highlighting these can capture cross-sectoral and cross-regional interactions that are missed by other approaches. They discuss the challenges, but also suggest ways to improve adaptation in these models through interdisciplinary studies and stakeholder and decision maker's discussions.

The papers by Street and also by van den Hurk look at the role of information in adaptation. Street sets out some of the issues for the application of climate services to support adaptation modelling, recognising the growing demand for such services and the role they will play in accelerating adaptation in Europe. He highlights a need to develop and focus products and services that directly inform and support solutions and actions, i.e. that make the transition to more applied adaptation. Van den Hurk looks at impact-oriented climate information selection, considering the challenges in formulating climate and socio-economic scenarios. He identifies the need for credible, relevant and legitimate information, with the use of tested models and concepts, but tailored to the decision context, and identifies one way to deliver this is through storyline development approaches that help select relevant and credible pathways to improve understanding of risks and adaptation options.

The next two papers focus on ways to improve current models and analysis. Wagener presents an evaluation of climate change impact models for adaptation decisions. He sets out that such models need to improve their analysis of local/regional implications to guide decisions makers and policy makers for adaptation strategies. He argues that impact models need to move beyond validation based on historical observations and use Global Sensitivity Analysis, i.e. to better understand the robustness of model outputs to input uncertainties over different projection horizons. Wilby recommends a greater focus on stress-testing adaptation options, i.e. that test the performance of adaptation projects despite uncertainty about climate change. This type of modelling analysis can help to identify conditions under which there may be trade-offs or even failure of project deliverables. These can be applied to projects, but can also be used for portfolios of options, using models of the system being managed. He also highlights the potential for field experiments and model simulations to test adaptation measures, as these provide information for real-world decision-making.

Finally, the paper by Ruiz Ramos and Rodríguez focuses on one of the key challenges, uncertainty, with an analysis of various techniques to reduce and manage uncertainty for adaptation. They provide a case study on handling the uncertainty of agricultural projections and adaptation, and highlight why these techniques can help increase adaptation effectiveness and avoid maladaptation.

Taken together, these papers provide conceptual thinking and new insights for adaptation modelling and they highlight areas and suggestions on how to overcome the various challenges needed to move to applied decision-making.

References

Chambwera M, Heal G, Dubeux C, Hallegatte S, Leclerc L, Markandya A, McCarl BA, Mechler R, Neumann JE (2014) Economics of adaptation. In: Field CB, Barros VR, Dokken DJ, Mach KJ, Mastrandrea MD, Bilir TE, Chatterjee M, Ebi KL, Estrada YO, Genova RC, Girma B, Kissel ES, Levy AN, MacCracken S, Mastrandrea PR, White LL (eds) Climate change 2014: impacts, adaptation, and vulnerability. Part A: global and sectoral aspects. Contribution of working group II to the Fifth assessment report of the intergovernmental panel on climate change. Cambridge University Press, Cambridge, United Kingdom and New York, NY, USA, pp 945–977

Klein RJT, Midgley GF, Preston BL, Alam M, Berkhout FGH, Dow K, Shaw MR (2014) Adaptation opportunities, constraints, and limits. In: Field CB, Barros VR, Dokken DJ, Mach KJ, Mastrandrea MD, Bilir TE, Chatterjee M, Ebi KL, Estrada YO, Genova RC, Girma B, Kissel ES, Levy AN, MacCracken S, Mastrandrea PR, White LL (eds) Climate change 2014: impacts, adaptation, and vulnerability. Part A: global and sectoral aspects. Contribution of working group II to the Fifth assessment report of the intergovernmental panel on climate change. Cambridge University Press, Cambridge, United Kingdom and New York, NY, USA, pp 899–943

Moser SC, Ekstrom J (2010) A framework to diagnose barriers to climate change adaptation. In: Proceedings of the national academy of sciences of the United States of America, vol 107(51). https://doi.org/10.1073/pnas.1007887107

OECD (2015) Climate change risks and adaptation: linking policy and economics. OECD Publishing, Paris. https://doi.org/10.1787/9789264234611-en

UNEP (2018) The adaptation gap report. Published by the United Nations environment programme, Nairobi. https://www.unep.org/resources/adaptation-gap-report-2018

Wilby RL, Dessai S (2010) Robust adaptation to climate change. Weather 65(7). https://doi.org/10.1002/wea.543

Chapter 1
Modelling the Cost and Benefits of Adaptation. A Targeted Review on Integrated Assessment Models with a Special Focus on Adaptation Modelling

Reimund Schwarze, Quirin Oberpriller, Martin Peter, and Jürg Füssler

Abstract This paper gives a targeted review on Integrated Assessment Models (IAMs) with a focus on damage functions and adaptation modelling.

Keywords Adaptation modelling · IAMs

Introduction

Integrated Assessment Models (IAMs) can be roughly distinguished according to their objective as cross-sectoral optimization models (POMs) and economic assessment of climate policies (PEMs) according to Toth (2005). The first are welfare-oriented optimal growth or general equilibrium models; the second are numerical simulation models for the cost minimization of climate policies in a partial or—also—a general equilibrium framework. The most popular models can be grouped according to this structure as in Table 1.1.

An evolutionary "family research" of IAMs helps to identify "generations" of models, where each generation enhanced the scientific understanding of climate change and influenced the economic recommendations for climate policy based on these models.

R. Schwarze (✉)
Helmholtz-Centre for Environmental Research–UFZ, Leipzig, Germany
e-mail: reimund.schwarze@ufz.de

Q. Oberpriller · M. Peter · J. Füssler
INFRAS–Forschung und Beratung, Zuerich, Switzerland
e-mail: quirin.oberpriller@infras.ch

M. Peter
e-mail: martin.peter@infras.ch

J. Füssler
e-mail: juerg.fuessler@infras.ch

C. Kondrup et al. (eds.), *Climate Adaptation Modelling*, Springer Climate,
https://doi.org/10.1007/978-3-030-86211-4_1

Table 1.1 Types of IAMs

IAMS: Integrated Assessment Models						
POMs: *Cross-sectoral* policy optimization				**PEMs:** Policy Evaluation		
Welfare optimization		**CGE**: General Equilibrium		**PE**: Partial Eq.	*Cost minimization*	
	Optimal Growth			Numerical Simulation		
	FUND, …	**DICE, RICE, ENTICE**, WITCH, MIND, …	**MERGE,** WIAGEM	Mini-CAM, **LPJmL, CAPRI**	**PAGE-2009**	*MESSAGE*

The first generation of models was developed in the 1990s as "basic models" that integrated climate modules and economic modules in one model. They all went through a period of "learning by doing" than "learning by investment", in short, they adsorbed endogenous growth theories that improved our understanding of the economic opportunities of a stringent climate policy, and ultimately shifted the 'optimal' or 'cost-efficient' policy paths towards demanding more and faster action against climate change. It was followed by a phase of learning on adaptation, often blended with geo-engineering which is the focus of this paper.

Effect-wise more important than all earlier are the two most recent generations of IAMs, which introduced climate catastrophes, economically speaking "fat tails" of climate risks, and the trend toward MIMI models, i.e. modularized, open source-based IAMs, which aim to overcome the popular criticism of a lack of transparency and black box-approaches of IAMs.

Figure 1.1 illustrates that not all model families have survived this process of "aging" and "learning", those that survived, re-acted to the scientific and societal critic by transformation.

Adaptation modelling challenges ahead.

Climate change impacts can be lowered by a variety of sectoral, regional and local adaptation measures. Including those in the damage function is a complex task for the following reasons:

- There are many climate-sensitive sectors each of which has specific adaptation measures, with respective costs and benefits in the short term and the long term.Especially for high temperatures, the benefits of adaption are highly uncertain (Fankhauser 2017).
- The extent and success of adaptation depends on the vulnerabilities and capabilities of regions and societies. Consider the example of the Netherlands and Bangladesh: Both will be highly affected by sea level rise, but the Netherlands is more able to handle such consequences, as the country is richer and has a long tradition of building sophisticated dikes. Damages will rise steeply if the adaptation capabilities of the affected societies are exceeded (Klein et al. 2008).
- The extent and form of adaptation is a choice by individuals and society. These choices may be modelled using a cost–benefit approach. Yet, information on

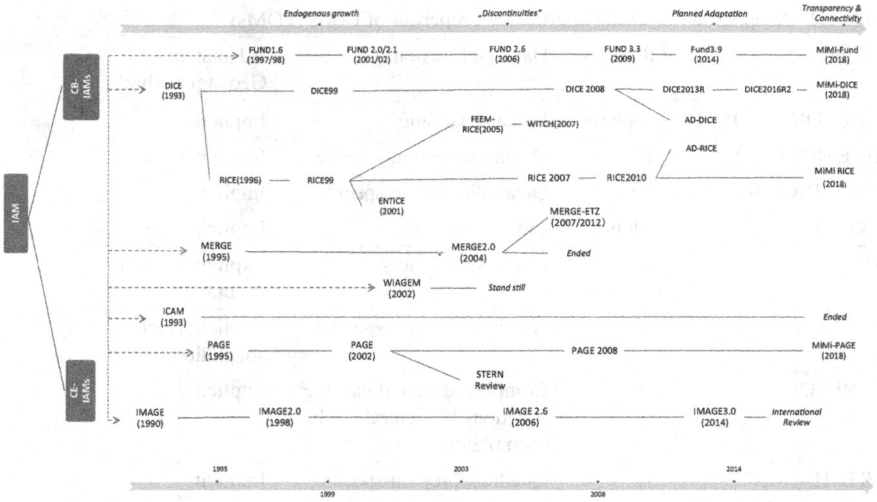

Fig. 1.1 History of IAMs

adaptation costs is scarce and local damages are uncertain. The literature on the costs and benefits of adaptation mainly considers coastal areas and agriculture.

IAMs dealing with adaptation.
Adaptation is incorporated in IAMs in very different ways (cp. Table 1.2):

- DICE considers adaptation implicitly. That is, the aggregate damage already includes the costs and benefits of adaptation (it is a "net" aggregate damage function). AD-DICE (de Bruin et al. 2009a, b) is an extension to DICE that explicitly considers adaptation. It disaggregates the damage function into adaptation costs and residual damages and selects a preferred combination of mitigation and adaptation.
- FUND introduces adaption for certain sectors explicitly. It includes an explicit cost–benefit analysis of costly coastal protection against sea level rise and assumes that parts of the agricultural damages (associated with the rate of climate change) fade with time at zero costs (autonomous adaptation). For other sectors, adaptation is implicit as in DICE (Diaz and Moore 2017; Estrada et al. 2019).
- PAGE introduces a tolerable temperature that increases with costly adaption measures. Damages are a function of the difference between the real and the tolerable temperature, such that, e.g. a real temperature increase of one degree without adaption causes the same damages as a real temperature increase of three degrees in case adaption has risen the tolerable temperature to two degrees.
- ICAM, MERGE and IMAGE consider geoengineering as an extreme form of adaptation, but at a different degree: some only as the carbon capture and storage (CCS), some also with CO2 removal (CDR) in different forms. To our knowledge, there is no IAM that would consider solar radiation management or ocean fertilization.

Table 1.2 Adaptation modelling approaches properties of IAMs (POMs)

Model	Regions	Damage function	Adaptation / Geoengineering
DICE (DICE- 2013R)	Global	Linear-quadratic	Implicit
RICE (RICE- 2010)	12	Quadratic, region-specific	Implicit
FEEM-RICE 10	10	Quadratic, region-specific	Implicit
ENTICE	Global	Linear-quadratic	Implicit
AD-DICE	Global	Linear-quadratic	Explicit; reactive adaptation
AD-RICE	12	Quadratic, region-specific	Explicit; reactive adaptation
FUND 3.9	16	Complex, different damage functions (8 sectoral DF for each region)	Implicit
WITCH	12	Quadratic, region-specific	Implicit
MERGE-ETL	Flexible	Quadratic, considers catastrophes	Implicit/Geo-Engineering (CCS, CDR)

Note CB-IAM = Cost–Benefit Integrated Assessment Model CB-IAM weighs mitigation costs against the benefits of reducing climate damages within one model. POM = Policy-optimizing models or fully integrated IAMs (Toth 2005)

There are similar differentiations of concepts of adaptation in PEMs.
Properties of IAMs (PEMs).

Model	Regions	Damage function	Adaptation / Geoengineering
ICAM	17	Complex	Implicit/Geo-Engineering (CDR, SRM)
PAGE09	8	Power function with uncertain exponent; considers catastrophes	Explicit; pro-active adaptation
IMAGE 3.0	26	Complex, biopysical feedbacks	Implicit (adaptation potential); soft-linking to GLOFRIS/FAIR / Geo-Engineering (CCS, CDR)

Note CE-IAM only assesses mitigation costs for a predefined mitigation target. More specifically, it calculates the most cost-effective (i.e. least-cost) way. PEM = Policy-evaluation models (Toth 2005).

- Adaptation measures can be further separated into measures that act quickly (e.g. air conditioning) as well as precautionary measures (usually infrastructure with a longlife-span) as in Fankhauser (2017). The latter is aimed at average climate change (better insulation of houses against the increase in summer temperatures) or at protecting against extreme events (e.g. dikes against floods).
- Auffhammer (2018) defines extensive and intensive margin adaptation. The extensive margin response is due to the installation of new equipment (e.g. new air

conditioning systems, irrigation equipment, sea walls, etc.). The intensive margin response means that existing equipment is used more frequently (the more frequent operation of existing air conditioners and irrigation equipment).
- Finally, the IPCC differentiates between adjustment costs (short-term costs of adaptation) and macro-scale adaptation (long-term restructuring of the economy). To correctly model the costs and benefits of adaptation all those different forms of adaptation have to be taken into account.

A special difficulty arises because adaptation costs can be seen as indirect damage costs. IAMs thus often blur the difference between direct damages (e.g. destructions caused by storms) and adaptation costs. In FUND, for example, the increasing energy cost of air conditioning is major damage sector, even though strictly speaking this is an adaptation measure. The corresponding decrease in damages (improved health) is not considered in FUND, even though a health sector exists. This obviously leads to an underestimation of the benefits of adaptation.

Another difficulty arises from the fact that the capacity for adaptation is a main defining element, and thus already explicitly considered for the SSPs. For example, in SSP1 and SSP5 the capacity to adapt is high, as there is a well-educated, rich population and a high development of technologies. In SSP1, there is in addition a good global governance and an intact ecosystem. In SSP3 and SSP4, on the other hand, the capacity is low due to the large, poor population, the lack of global cooperation, a slow technological development and unequal distribution of resources. These features have not yet been included in the damage functions in a harmonized manner.

Conclusions and Recommendation

To summarize, IAMs include adaptation explicitly (i.e. conducting a cost–benefit analysis of adaptation measures), implicitly (i.e. damage function is net of adaptation) or occurring autonomously (impacts fade at zero cost). In any case, these are highly aggregated approaches that do not consider the variety of adaptation possibilities at the local, regional and sectoral level. If at all, IAMs make very rough and ad-hoc assumptions on adaptation costs and benefits and do not include technological details. The understanding of (future) adaptive capacity, particularly in developing countries, through IAMs is still limited (Watkiss 2011). The extent and success of adaptation depends on the vulnerabilities and capabilities of regions and societies. In the face of all these shortcomings our recommendation, however, is not to give up on IAMs but to go through another phase of "aging and learning" of adaptation models that better fit to the heterogeneity of adaptation measures.

References

Aufhammer M (2018) Climate adaptive response estimation: short and long run impacts of climate change on residential electricity and natural gas consumption using big data. NBER Working Paper No. w24397. Available at SSRN: https://ssrn.com/abstract=3138381

Ciscar JC, Goodess C, Christensen O, Iglesias A, Garrote L, Moneo M, Quiroga S, Feyen L, Dankers R, Nicholls R, Richards J (2009) Climate change impacts in Europe. Final report of the PESETA research project.

De Bruin KC, Dellink RB, Tol RSJ (2009a) AD-DICE: an implementation of adaptation in the DICE model. Clim Change 95:63–81

De Bruin KC, Dellink RB, Agrawala S (2009b) Economic aspects of adaptation to climate change: Integrated assessment modelling of adaptation costs and benefits, OECD Environment Working Papers 6

Diaz D, Moore F (2017) Quantifying the economic risks of climate change. Nat Clim Chang 7(774–782):11

Dumas P, Hà Dương M (2008) Optimal growth with adaptation to climate change. HAL, Working Papers. 117. https://doi.org/10.1007/s10584-012-0601-7

Estrada F, Tol RSJ, Botzen WJ (2019) Extending integrated assessment models′ damage functions to include adaptation and dynamic sensitivity. Environ Model Softw 121. https://doi.org/10.1016/j.envsoft.2019.104504

Fankhauser S (2017) Adaptation to Climate Change. Ann Rev Resour Econ 9. https://doi.org/10.1146/annurev-resource-100516-033554

Klein RJT, Kartha S, Persson A, Watkiss P, Ackerman F, Downing TE, Kjellén B, Schipper L (2008) Adaptation: Needs, Financing and Institutions. Breaking the climate deadlock briefing paper. www.breakingtheclimatedeadlock.com

Oberpriller Q, Peter M, Füssler J, Zimmer A, Schaeffer M, Aboumahboub T, Schleypen J, Roming J, Schwarze R (2021) Climate cost modelling—analysis of damage and mitigation frameworks and guidance for political use. Report for the German environment agency (forthcoming)

Ortiz R, Markandya A (2010) Literature review of integrated impact assessment models of climate change with emphasis on damage function. BC3 Working Papers 6, Bilbao

Toth FL (2005) Coupling climate and economic dynamics: recent achievements and unresolved problems. In: Haurie A, Viguier L (eds) The coupling of climate and economic dynamics. Advances in global change research 22. Springer, Dordrecht

Watkiss P (2011) Aggregate economic measures of climate change damages: explaining the differences and implications. Wires Clim Change 2011(2):356–372

Chapter 2
Cross-Sectoral Challenges for Adaptation Modelling

Asbjørn Aaheim, Anton Orlov, and Jana Sillmann

Abstract Socioeconomic studies on adaptation based on bottom-up approaches have been focusing mainly on local impacts of weather-related variations, thereby neglecting potential remote impacts. There is little knowledge about challenges that relate to the global and long-term character of climate change. By contrast, impact assessment studies using top-down approaches, such as multi-region, multi-sector computable general equilibrium (CGE) models, provide a consistent framework to capture potential remote impacts, which occur through cross-sectoral and cross-regional interactions. Here we present main findings of our economic impact assessments of climate change and adaption modelling. Furthermore, we discuss the challenges for incorporating adaptation measures and policies into macroeconomic models.

Keywords Adaptation · Impacts of climate change · Macroeconomic modelling · Cross-sectoral interactions

Introduction

Most economic studies on climate adaptation focus primarily on local impacts of climate change on people, companies, and local authorities, whereas potential remote impacts are typically not considered. Attempts to consistently integrate lessons from studies of impacts and adaptation in a model that addresses cross-sectoral impacts indicate how important interdependencies across sectors are, but they also reveal knowledge gaps related to implementing measures to motivate adaptation across levels. Economic theory provides a formal link between descriptions of individual behaviour and economic drivers on the national level. This is the idea behind the macroeconomic model GRACE, which is based on a standard description of relationships between economic activities in countries and describes how they depend on the activities in other countries (Aaheim et al. 2018). The framework of GRACE enables

A. Aaheim · A. Orlov · J. Sillmann (✉)
CICERO Center for International Climate Research, Oslo, Norway
e-mail: jana.sillmann@cicero.oslo.no

C. Kondrup et al. (eds.), *Climate Adaptation Modelling*, Springer Climate,
https://doi.org/10.1007/978-3-030-86211-4_2

to depict cross-sectoral and cross-regional interactions and to integrate lessons from the research on adaptation to project economic development. Moreover, the flexible and consistent modelling framework of GRACE can be used to assess the trade-offs between climate change mitigation and adaptation policies. In the following, we briefly describe the GRACE model as adaptation modelling tool (Sect. "The Grace Model") and then give several examples of studies where the GRACE model is applied to different sectors (Sect. "Implications of National and Global Dependencies in Adaptation"). In Sect. "Challenges for Adaptation Modelling", we discuss challenges and opportunities for complementing bottom-up with top-down approaches in adaptation modelling.

The Grace Model

GRACE is a global computable general equilibrium model, which uses data from the national accounts, collected by GTAP (Global Trade Analysis Project) data base (Aguiar et al. 2019). The data give values of deliveries from production sectors to other production sectors and to the final consumption. These are illustrated by the two upper grey boxes in Fig. 2.1. Production of goods and services require input of intermediates, labour, capital, and natural resources. Total input in each production sector as well as the composite of goods for consumption and investments can be read from the columns (red line), while the rows divide total demand for each sector product on other sectors and final deliveries (green line). The demand for each good is divided into domestic demand and demand from the other regions. The regional and sectoral aggregation is flexible depending on the scope and focus of a study. The GTAP database is used to calibrate the demand and production systems in the model. They thereby provide supply and demand functions (curves on the right), and the model finds the combination of prices and quantities under the assumption of market

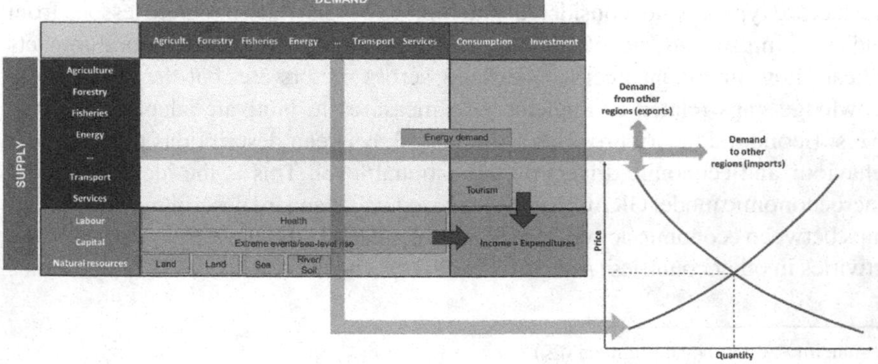

Fig. 2.1 Structure of GRACE

equilibrium, where supply equals demand. Economic growth is driven by underlying changes in population (labour force), capital (investments and technological change), availability of natural resources and impacts of climate change.

There are two main types of adaptation in GRACE related to i) adaptation in production and ii) adaptation in consumption. The former one refers to the substitutability among production inputs (e.g., labour, capital, and natural resources, including land) to achieve a certain level of production. Furthermore, cross-regional and cross-sectoral mobility of labour and capital are another channel of adaptation in production. The latter one deals with the substitutability between imported and domestically produced goods and services as well as substitutability among different commodities in final consumption.

Implications of National and Global Dependencies in Adaptation

GRACE provides an opportunity to utilize knowledge on mitigation options and physical impacts of climate change to assess the global consequences on standardized, economic indicators for evaluation of the economic development in countries and regions. The GRACE model confirms that without mitigation policies, the impacts of climate change on economies can be large, which implies high costs of adaptation. For instance, Aaheim et al. (2017b) addressed the economic consequences of mitigating climate change by reducing emissions from a high-emission pathway that leads to an increase in global mean temperature between 5.0 and 6.0 °C (RCP8.5) to approximately 3.0 °C in a moderate emission pathway (RCP4.5) in 2100. The mitigation alternative (i.e., RCP4.5) gives moderate costs in some regions and small benefits in others. A high-emission pathway (i.e., RCP8.5), on the other hand, implies huge costs in all regions, occurring mainly after 2070. The explanation to the moderate impacts in both alternatives in the coming decades is adaptation among economic agents with resulting price effects with impacts on sector composites and trade. The potential for this adaptation is limited, however, as becomes apparent under stronger impacts associated with high emissions.

The study shows that RCP4.5 may represent an optimal, global strategy, as it leads to a stabilization of climate change impacts in the long term. At the same time, it illustrates the challenges in achieving the "best solution for future generations", as the policy needed to achieve a 3.0 °C target implies an increase in the cost of emitting a ton of CO_2 at 12—15 percent per year from 2010 to 2075, according to this study. This is far beyond the shadow costs of cutting emission today, despite the ambitions in the Paris Agreement. The model is further developed with the aim of specifying economic activities that will be affected directly by climate measures or impacts of climate change. Some examples are given below.

Table 2.1 Impacts of climate change (in Bill US$) expected in 2100 on the value added from forests world-wide with and without adaptation in three emission scenarios (RCPs, van Vuuren et al. 2011) and three adaptation scenarios

	RCP2.6	RCP4.5	RCP8.5
No adaption	−1.8	−1.5	−2.5
Adaption in forest management	7.5	5.2	8.0
Adaption in management and harvesting	28.7	17.4	26.4

Adaptation in the Forestry Sector and the Global Impacts

Aaheim and Wei (2020) estimate the economic consequences of climate change on the biomass in forests in 27 countries in the world under different emission scenarios based on the physical estimates, which included adaptation in the utilization of forests and resulting market effects. They use the GRACE model with a modified module for forest management, where the standing biomass is interpreted as wealth, similar to the stock of capital, and the harvesting is interpreted as the return of this wealth. The estimates were further presented with three alternative scenarios of "economic impacts" (see Table 2.1). The first scenario reveals the economic cost of the impact on the growth of the biomass, if there are no price effects or any adaptation. The second scenario shows the impact on the value of harvested wood after adaptation to the new relationship between the growth and stock of biomass and to the market effects of the impacts in world markets. The third scenario shows the impact on the value added in the forestry sectors, where adaptation takes place also in the harvesting of forests, reflected by the production functions in the model. Examples are to leave less wood for waste, to improve the transport network or to harvest less accessible parts within an area.

If based on the physical assessment alone, climate change leads to a reduction in the value added from forests world-wide. This is partly due to the variations in value added across regions, where the growth of forests is negatively affected in regions with a high value added. Under the 'No adaptation' scenario, the physical effects on forests are negative in all RCPs. Price effects and resulting adaptation related to the management turns the impacts on the value of forests positive, however. In addition, adaptation related to the harvesting implies that the forestry sectors will benefit from getting more out of each m^3 harvested. This further increases the positive economic impacts. It must be noted that the impacts vary a lot across regions and in the different RCPs. In some regions, the value of forests decreases under RCP2.6 and RCP4.5.

Adaptation from a Local Perspective and the National Impacts

In general, there are two approaches to assess the impacts of climate change, such as bottom-up and top-down. The former one addresses local impacts on households, farmers, companies, and local authorities. The latter one deals with a broader (macroeconomic) implications of climate impacts. Both approaches have their strengths and weaknesses. Aaheim et al. (2018) provided a consistency check in assessments of economic impacts of climate change on aggregated levels with local studies of impacts and adaptation among smallholders in Nepal up to 2050 under a high emission pathway (RCP8.5). The impact to the Nepalese economy is assessed with the basic version of GRACE, where the impacts to agriculture are quantified based on a meta study of estimates from different integrated assessment models. It is assumed that the productivity of land will be reduced by 2.7 percent in 2050 on average for Nepal.

To address the vulnerability of smallholders, the study also includes an assessment of the impacts to farmers in a local community. Descriptions of the sources of livelihood among smallholders (e.g., consumption of own produced goods, products bought in market and work time spent on own farm and outside the farm) were collected from interviews of 60 households in Bamrang Khola village in Khotang district of Nepal (Aaheim el al. 2017a). The data were used to analyse how the economic behaviour described on the macro level depends on constraints that are ignored in GRACE. First, the output is constrained by the size of the farm. Second, there is a division between food products sold in markets and food for the farmers' own consumption. Third, farmers can work outside the farm to gain monetary income. Hence, their consumption is partly based on market transactions.

The impacts of climate change refer to the projection for Bamrang Khola in the climate projection for Nepal used in the macroeconomic analysis indicating a reduction in the productivity of land at two percent and health effects, which reduces the productivity of work by 2.4 percent. Because of the inability to take full advantage of the price effects indicated by the macroeconomic analysis, the study of the impacts to the local households gives a very different picture of the vulnerability of people in the agricultural sector in Nepal than what comes out of the macroeconomic analysis. The impacts differ also depending on the size of the farms disposed by the different households.

The message from this study is, firstly, that the constraints related to the farm size puts strong limitations to adaptation in most of the farms. Most of the households are subject to constraints that enforce a notable reduction in work on the farm, which can be compensated only by a slight increase in monetary income needed to compensate losses in the production of food from farm with food and other goods from the market. These impacts are disregarded in the macroeconomic assessment. The second message is the difference between the estimated impacts on the Nepalese economy in the macroeconomic assessment and the impacts to households in the micro-based assessment, despite the attempt to base both on the same description of the future. While GDP and consumption are reduced by approximately 0.25 percent in Nepal according to the macroeconomic analysis, sources of livelihood for most of the smallholders are reduced by more than two percent, depending on the size of the farm, according to the household study. There are possible explanations to this, which need to be confirmed by further studies. This points to the need for a better understanding of how climate change affect parts of societies, and the most vulnerable, to coordinate targeted adaptation strategies across levels.

Adaptation to Heat Stress

Orlov et al. (2020) conducted an economic impact assessment of heat-induced impacts on worker productivity under RCP2.6 and RCP8.5. The authors implement autonomous adaptation in the GRACE model, such as penetration of air conditioners and mechanization of outdoor work in agriculture and construction, which were linked to different SSPs. Projections of mechanization for different SSPs were extrapolated based on a regression analysis, where the number of tractors per hectare was used as a proxy for mechanization and GDP per capita as an explanatory variable. Results from model simulations showed that under RCP8.5 by 2100, heat-induced reductions in worker productivity lead to an average decline of 1.4% in GDP relative to the reference scenario with no climate change (see Fig. 2.2). This is approximately 0.4 percentage points less than when no autonomous mechanization is assumed.

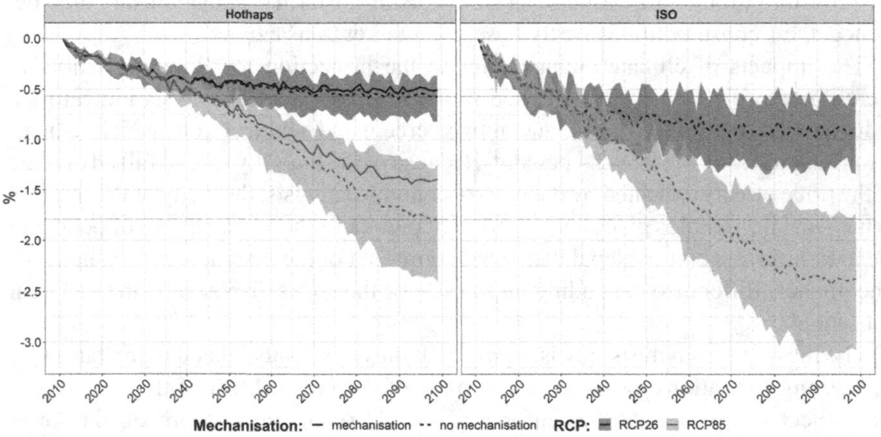

Fig. 2.2 Changes in global GDP under RCP2.6 and RCP8.5 compared to the reference scenario with no climate change. The solid and dashed lines show the mean values and the shaded areas indicate uncertainties in the estimated impacts. In the plot legend, "mechanization" (solid lines) stands for decreasing work intensity due to mechanization driven by an economic growth, while "no mechanization" (dashed lines) implies a constant work intensity. "Hothaps" stands for the epidemiological exposure–response function, while "ISO" implies that ISO: 7243:1989 standards are used to assess heat stress impacts on worker productivity

Challenges for Adaptation Modelling

Drawing from the results of above presented studies, we emphasise the importance of adaption measures and policies to deal with adverse climate-induced impacts. We find that adaptation plays a key role in diminishing adverse impacts of climate change. Furthermore, we identify several challenges for incorporating adaptation into global macroeconomic models. While most of local studies on adaptation using bottom-up modelling approaches neglect potential remote impacts, top-down economic models, such as CGE models, feature great macroeconomic consistency and can capture cross-sectoral and cross-regional interactions. However, due to modelling complexity, multi-region multi-sector macroeconomic models are very aggregated and underly simplified assumptions. Often, market-related barriers and inertia are not implemented in those models, which might lead to an underestimation of economic costs arising from climate change. The studies presented above indicate that the impacts to countries and to national economies may differ substantially from the conclusions drawn from bottom-up approaches. On the other hand, the generalizations used to address impacts and adaptation on a national level in top-down approaches have a weak empirical foundation in most cases. The examples above show the potential for improving the linkages, for instance, by informing the top-down modelling by refining the modelling assumptions or damage functions with local insights from interviews or field experiments. There is a need for a better integration of results and methodology from local studies on climate impacts and adaptation into a broader framework of macroeconomic models to provide more consistent and accurate impact assessments. And vice versa, local studies based on bottom-up methods might greatly benefit from insights obtained from top-down analyses, which can set some boundary conditions including indirect cross-sectoral and cross-regional economic responses. Therefore, there is a big potential in the complementarity of top-down and bottom-up approaches, which is not yet fully realised. To bridge the top-down/bottom-up divide in adaptation modelling, more interdisciplinary studies involving the combination of two approaches are needed. Also, the involvement of stakeholders and decision-makers is vital to better understand challenges for adaptation planning in practice.

Moreover, many adaptation measures are implemented autonomously in CGE models, which means that the implementation of adaptation measures depends on the assumed socioeconomic scenarios (e.g., SSPs), and these are not a part of optimisation in the model (i.e., no proactive adaptation). Adequate representation of extreme events in models based on national aggregates with annual data, and the sensitivity to the variability of impacts within countries in the national aggregates can further improve CGE modelling. Finally, most of the large-scale CGE models applied for impact assessments of climate change are either static or recursive-dynamic, which implies that consumers and producers make their decisions based on current prices, whereas anticipatory adaptation might play a big role in reducing the future economic cost of climate change. Therefore, a better integration of proactive adaptation under uncertainties into macroeconomic models is needed.

Acknowledgements The authors are supported by the European Union's Horizon 2020 research and innovation programme under grant agreement No 820655 (EXHAUSTION) and No 820712 (RECEIPT).

References

Aaheim A, Orlov A, Dhakal K (2017a) Socio-economic impacts of climate change in Hindu-Kush Himalaya, CICERO Report 2017:07. CICERO Oslo

Aaheim A, Wei T, Romstad B (2017b) Conflicts of economic interests by limiting global warming to +3 °C. Mitig Adapt Strat Glob Change 22(8):1131–1148. https://doi.org/10.1007/s11027-016-9718-8

Aaheim A, Orlov A, Wei T, Glomsrød S (2018) GRACE model and application, CICERO Report 2018:1. CICERO, Oslo

Aaheim A, Wei T (2020) Relationships between physical effects of climate change on forests and economic impacts by world region, CICERO Report 2020:02. CICERO, Oslo

Aguiar A, Chepeliev M, Corong EL, McDougall R, van der Mensbrugghe D (2019) The GTAP data base: Verions 10, J Glob Econ Anal 4(1):1–27. https://doi.org/10.21642/JGEA.040101AF

Orlov A, Sillmann J, Aunan K, Kjellstrom T, Aaheim A (2020) Economic costs of heat-induced reductions in worker productivity due to global warming. Glob Environ Chang 63. https://doi.org/10.1016/j.gloenvcha.2020.102087

van Vuuren DP, Edmonds J, Kainuma M et al (2011) The representative concentration pathways: an overview. Clim Change 109:5. https://doi.org/10.1007/s10584-011-0148-z

Chapter 3
Climate Services Supporting Adaptation Modelling

Roger B. Street

Abstract Information and knowledge resources to support climate action (climate services) have been the subject of investments by European and national funding agencies and by the private sector in response to a growing demand and requirements to support climate-related actions. The extent to which the current state of these resources are consistent with and attributable to these investments still requires further assessment. There have been efforts to continue to inform such investments and to stimulate the climate service market. These to some degree identified remaining and emerging gaps, including those intending to support enhancing the breadth, quality and relevance of products and services, the infrastructure supporting the climate service public and private market domains and the factors enabling market growth. The criticality of realising the benefits from the availability and use of this intelligence is increasing and evolving as Europe and the rest of the world call for a transition to a climate-resilient and a low-carbon society and economy. To realise and sustain this potential, there is the need for a systematic assessment of the impacts of previous investments and of where and what type of investments could enhance the impacts in terms of informing action—exploring and identifying shared pathways to enable the development and use of climate services.

Keywords Climate services · Intelligence to inform solutions · Investment gaps

The European Research and Innovation Roadmap for Climate Services

Rather than setting a definitive research and innovation agenda, this Roadmap (EC (RTD) 2015) offers a framework to engender discussions among relevant actors and stakeholders, thereby providing a means of exploring and finding shared solutions and pathways that enable the development and use of climate services that provide

R. B. Street (✉)
Environmental Change Institute, University of Oxford, Oxford, UK
e-mail: roger.street@ouce.ox.ac.uk

C. Kondrup et al. (eds.), *Climate Adaptation Modelling*, Springer Climate,
https://doi.org/10.1007/978-3-030-86211-4_3

benefits to society. In so doing, the intention of such was that these benefits included informing solutions (climate action).

For the purposes of the Roadmap, climate services were defined as follows:

> '…covering the transformation of climate-related data—together with other relevant information—into customised products such as projections, forecasts, information, trends, economic analyses, assessments (including technology assessments), counselling on best practices, development and evaluation of solutions and any other service in relation to climate that may be of use for the society at large.'

As such, these services were seen as including data and information, reflect knowledge that support adaptation, mitigation and disaster risk management and, as such, could be described as comprised of four interrelated components as follows:

- Guidance informing climate adaptation and resilience journeys, often outlined in terms of informing risk assessment or risk management and adaptation planning and implementation frameworks;
- Data and information to support these assessments and processes (including climate and socio-economic data and information);
- Tools and resources to support these assessments and processes; and
- Enablers including knowledge brokering, capability development, use journeys, case studies, training/capacity development, communication activities and communities of practice.

In this sense, they are seen as offering a range of services and products intended to better inform decision-making and their implementation by the spectrum of decision and policy makers—within the public sector across all levels of government, within communities, within civil society, businesses and industries, and for individuals. By doing so, climate services are seen as having the potential to become ever-green enablers of smarter, systemic and timely climate action.

What Has Been Accomplished and What Are Still Seen as Gaps?

The Roadmap identified three research and innovation challenges comprised of nine main activities and 25 specific actions, addressing that which could facilitate the development of a climate service market capable of enabling and empowering climate action".

1. **Enabling market growth**—assessing the nature, growing and demonstrating the added value of a climate service market;
2. **Building the market framework**—supportive communities and infrastructure; standards, quality assurance and control, access and legal aspects; and international cooperation; and

3. **Enhancing the quality and relevance** of climate services—supportive information frameworks, strengthen the scientific basis and relevance; and climate information and users' needs, innovations and products.

Over the past five years there have been some reviews and shared perspectives on research and innovation gaps undertaken under various guises. The following provides highlights of the results of three of these.

Following a stakeholder engagement workshop (November 2017) held at the request of the European Commission (linked to the informal DG RTD European Climate Services Roadmap Implementation Working Group), a position paper was prepared by Climateurope (2018) summarising deliberations in the form of recommendations for climate services science, research and innovation. The main messages and advice were grouped under the following:

- Setting standards, quality control, quality assurance and evaluation [Linked to Roadmap Challenge 2].
- Legal and ethical considerations [Linked to Roadmap Challenge 2].
- Authoritative voice [Linked to Roadmap Challenge 2].
- Knowledge transfer/brokering, communication and user engagement [Linked to Roadmap Challenge 2].
- Capacity building [Linked to Roadmap Challenge 2].
- Mainstreaming climate services [Linked to Roadmap Challenge 2 and Challenge 1].
- Balance between underpinning science, research, innovation and market growth [balance investments across the three Roadmap challenges].
- Sustainable climate data sources [Linked to Roadmap Challenge 2].

As indicated, the resulting stakeholder assessment is consistent with aspects of the Roadmap challenges and activities. The apparent focus of the identified gaps suggests that at that stage in time the engaged stakeholders (users, providers and researchers) believed there was a need to continue efforts focusing on building the market framework and also recognised the need to retain an appropriate balance in investments that continued supporting the underpinning research and associated infrastructure, in addition to supporting innovations. The highlighting of this latter area was to some degree the result of concerns regarding the increasing focus on innovations.

Climateurope (2019) provides recommendations for the Horizon Europe (HEUR) framework programme on research needs for climate modelling and climate services. These recommendations were elaborated within Climateurope by a group of European experts in climate modelling and climate services and were prepared under the coordination of a small group of scientists from various European research institutions participating in the project. Recommendations directly related to climate services were presented as follows:

- Supporting the formulation of adaptation strategies—systematic availability of impact-oriented projections and up-to-date near-term predictions downscaled to local scale, process understanding, models and infrastructure, downscaling, along with guidance to support selection, aggregation, and use of the local climate information. [Linked to Roadmap Challenge 3].

- Understanding requirements, decision-making context and foresight for climate services—research should contribute to increasing knowledge towards reaching societal goals including by triggering cross-pollination between social and natural sciences to include the human dimension into climate services research. [Linked to Roadmap Challenge 1].
- Enhancing diffusion of innovation and information for climate—operationalising climate services to facilitate adoption of innovative practices that support adaptation today and in the longer-term [Linked to Roadmap Challenge 1].
- Assessing the value of climate services—better understanding of the underlying values (expected and potential ecological, social, ethical and economic value) and why some undervalue climate services is needed to increase the pull for climate services. [Linked to Roadmap Challenge 1].
- Standardizing climate services—to generate trust across supply and demand there is a need for a coherent and agreed upon set of authoritative standards for the overall value chain. [Linked to Roadmap Challenge 2].
- Strengthening the links between the climate modelling and climate service communities—benefit for both communities in term of informing and rationalising the pull for outputs from climate modelling and impact communities' activities and informing the potential for additional (and potentially more) relevant climate services based on research directions and outputs. [Linked to Roadmap Challenge 2].

As suggested, these recommendations are consistent with and build on the challenges and activities comprising the Roadmap. There breadth reflects the interests and perspectives of the community engaged in developing and articulating this set of recommendations. One point of interest is that this paper was developed for the HEUR cluster identified as 'climate science and solutions' and the recommendations reflect to some degree the emerging focus on 'solutions' within the European science and policy communities.

A workshop held 09–10th June 2020 under the auspices of the JPI Climate ERA4CS to support the further implementation of the JPI Climate Strategic Research and Innovation Agenda (SRIA) included consideration of future research needs in support of climate services. Among the challenges highlighted were as follows:

- Many initiatives supported by research projects stop before reaching the stage of providing a stand-alone, operational service, even when they have been successful in meeting user requirements. [linked to the need to operationalise climate services].
- The majority of potential users are not yet convinced of the value of climate services [Linked to Roadmap Challenge 1].
- Users do not necessarily see climate change as a standalone risk and prefer an integrated approach addressing all risks. [Linked to Roadmap Challenge 1].
- The development of CS is hampered by both scientific and communication difficulties.

On the climate science side:

- Seasonal and decadal predictions are still of insufficient quality to convince most users that climate prediction is a mature science. [Linked to Roadmap Challenge 3].
- Longer term (century) climate simulations suffer from a large dispersion of model results (e.g., future precipitation in Africa). [Linked to Roadmap Challenge 3].
- Climate models often have an insufficient resolution to deliver relevant results for users. [Linked to Roadmap Challenge 3].

On the communication side:

- CS still use terminology which is not understood and often misinterpreted by most users—need for standardised terminology and quality assurance. [Linked to Roadmap Challenge 2].
- Knowledge elements resulting from research could provide the basis for future climate services; however, realising this potential is challenging due to limits in translational capabilities, including across disciplines [Linked to Roadmap Challenge 3].

- There is a need for a 'platform' (or a network of 'platforms') to facilitate sharing experiences and lessons learnt and collaboration on mutually interesting challenges (enhance complementarities and minimise conflicts and duplication). [Linked to Roadmap Challenge 2].
- Need to shift from a focus on supporting incremental adaptation to also supporting transformational adaptation, including transitions. [Linked to Roadmap Challenge 1—understanding user needs].
- Lacking an overarching framework and metrics for evaluating climate services, including consideration of relevance, usability and legitimacy in addition to credibility [Linked to Roadmap Challenge 2].
- Exploring and identifying good practice business models [Linked to Roadmap Challenge 1].
- Enhanced level of inter- and trans-disciplinary in science supporting climate services [Linked to Roadmap Challenge 3].
- Coordination at the European level in scenario development, including use as well as development, would enhance adaptation and resilience considering interdependencies and the systemic nature of climate risks and solutions [Linked to Roadmap Challenge 2].

Conclusions and Recommendations

In 2015, the Roadmap recognised that there was a great potential for enhancing the climate service sector based on its perceived value to society. The demand for such services was seen as increasing but in need of encouragement including by increasing the marketability and quality of the products and services available (enhancing the service push) and used (enhancing the service pull). This understanding of the state of the market remains true today and is increasing. The breadth of the demand-side of the climate service market has broadened and deepened as a result of emerging policy requirements and understanding of the need for action, including the following:

- The climate-resilience and low-carbon focus within the European Green Deal and similar initiatives elsewhere;
- The ever-growing focus on a climate-resilient and low-carbon focus for COVID-19 recovery investment; and
- The increasing demands by the investment and banking sector related to supporting transition and physical risk management related to economic sustainability concerns.

The broadening and deepening of the demand are further evident considering the development of the proposed EU Horizon Europe missions, specifically 'Accelerating the transition to a climate-prepared and resilient Europe'. The mission board in presenting their proposal recognised the fundamental importance of climate services. It also has been suggested that without climate services, this and the other missions—Regenerating our Ocean and Waters; 100 Climate-Neutral cities by 2030—by and for the citizens; and Caring for Soil is caring for life—would be 'missions impossible'.

A particular aspect of this change in the market that warrants highlighting in the context of adaptation modelling is the increased focus on services supporting solutions and actions as evident in the European Green Deal and the EU Horizon Europe missions. As such, to be effective and relevant, climate services will need to specifically develop and focus products and services that directly inform and support solutions and actions.

Investments by identified actors, including Horizon 2020, in the context of responding to the challenges identified within the European Roadmap have had impacts. These impacts are evident in increased understanding of aspects of the supportive sciences, increases in the scope, relevance and accessibility of service and in efforts directed at facilitating the market pull. The extent to which these are attributable to the investments still requires further assessment. In addition, considering that the Roadmap was launched six years ago (2015), the scope and nature of investments made to address the identified challenges, the evolving nature of the climate service market and of the related policy and practice landscape, many believe that in addition to an assessment of impacts, there is also a need for a systematic assessment of remaining and emerging challenges.

The breadth of such an assessment should include addressing questions regarding future directions among which are as follows:

- To what extent are requirements for solution- and action-supportive climate services being reflected in the market (demand and supply)?
- To what extent are climate products and services currently available and under development able and recognised as being able to support and inform climate

actions (resilience, adaptation and mitigation) and related processes? Consider the perspectives of those providing and those using those services and products.

- To what extent are current research and innovations efforts/directions and related funding considering the need to support the evolution in climate services that evolving requirements are and will demand?
- Are innovations that could be directed at enhancing the relevance, usability, legitimacy and credibility of solution-based climate services and products consistent with what are and will be required to inform actions as reflected in the European Green Deal and EU Horizon Europe missions?

In addition, I would suggest there is an overarching process-based question that also requires consideration:

- Are processes and support mechanisms that are intended to facilitate the transition of research/project-created products and services to operations sufficient/effective? What could be done to further facilitate this transition?

References

Climateurope (2018) Position paper on recommendations for climate services science, research and innovation. https://ec.europa.eu/research/participants/documents/downloadPublic?documentIds=080166e5be3042e3&appId=PPGMS. (Accessed 15 Jul 2020)

Climateurope (2019) Recommendations to Horizon Europe on research needs for climate modelling and climate services. https://www.climateurope.eu/recommendations-to-horizon-europe-on-research-needs-for-climate-modelling-and-climate-services/. (Accessed 15 Jul 2020)

European Commission (DG RTD) (2015) A European research and innovation roadmap for climate services. https://op.europa.eu/en/publication-detail/-/publication/73d73b26-4a3c-4c55-bd50-54fd22752a39. (Accessed 15 Jul 2020)

Chapter 4
Impact-Oriented Climate Information Selection

Bart van den Hurk

Abstract To support climate adaptation decision-making, a picture of current and upcoming climate and socio-economic conditions is required, including an overview of intervention scenarios and their impact. In order to be actionable, this picture needs to rely on credible, relevant, and legitimate information, which implies the use of tested models and concepts, tailored to the decision context, and with transparent and understandable assumptions on boundary conditions and process representation. These criteria are challenged when the complexity of the problem is large and stakes are high. For many conditions, unforeseeable features and events with potentially large implications affect the problem at hand and contribute to the uncertainty that is not easily quantified, let alone eliminated. We explore storyline development approaches that help in selecting relevant and credible pathways and events that enrich the understanding of the risks and options at stake. We explore two categories of storylines (climate scenario storylines and climate risk storylines) by discussing use cases in which these were developed.

Keywords Climate information · Storylines · Climate adaptation · Decision support · Risk information

Introduction

The world is a complex place, and the predictability of its dynamics is further challenged by climate change and its myriad of impacts. To support adaptation to these potentially high-impact future conditions, a description of current and future conditions and options needs to be provided, that is built on credible, relevant, and legitimate information (Vincent et al. 2018). For this, a proper system definition needs to be described that allows quantitative evaluation of (adverse) impacts of hazards and determination of the probability of event cascades. For this, a wide range of

B. van den Hurk (✉)
Deltares, The Netherlands
e-mail: bart.vandenhurk@deltares.nl

C. Kondrup et al. (eds.), *Climate Adaptation Modelling*, Springer Climate,
https://doi.org/10.1007/978-3-030-86211-4_4

models is used to map potential pathways and situations, and the effects of adaptation interventions on these (Van den Hurk et al. 2018). However, credible predictions of future conditions are severely constrained by unknown drivers (greenhouse gas emissions, land use change, societal exposure, and vulnerability to hazards) and imperfect foresight capabilities (biased models, internal variability). In addition, complex compounding occurrences of drivers or impacts may lead to unforeseeable events or pathways that leave a large impact on the assessment of the current and future risk profile (Zscheischler et al. 2018). The higher the complexity of the situation, and the higher the stakes, the larger the challenge to meet the criteria of credibility, salience, and legitimacy.

To explore future conditions that are highly unpredictable but may unfold society-relevant impacts, a vast tradition of scenario construction has been developed over the past decades. In the field of climate change well-known benchmarking products are the emission and societal transition scenarios (RCPs, SSPs, Riahi et al. 2017), the modeled climate response and potential impacts (CMIP, CORDEX, ISIMIP, e.g. Eyring et al. 2016), and the expanding collection of national climate scenario and climate impact assessments embedded in regional, national, or European climate adaptation policy frameworks.

Storylines are a necessary element in these scenario frameworks. They provide a compelling and consistent narrative that is deemed plausible and relevant, and form the backbone logic of the scenarios that are derived from these. Storylines essentially consist of plausible assumptions on conditions and processes, and require the involvement of experts on these developments (Shepherd et al. 2018), both from practitioners and a scientific point of view. Many scenario frameworks provide multiple storylines, either to contrast potential but inconsistent storylines, or a form of uncertainty range to the collection of scenarios, in order to give an indication of the operation or tolerance range for which adaptation policies need to be designed.

Guided by a number of ongoing climate research programs we explore and contrast two types of climate storylines: climate scenario and climate risk storylines, both designed to condense the wide range of potential climate change projections into a compelling range that is relevant for impact assessment retaining as much as possible a link to the real world's experience with societal climate change impacts. The *climate scenario storylines* are used to aggregate a large volume of global climate change projections into a discrete set of stakeholder-oriented national scenarios. *Climate risk storylines* are mapping climate-related shocks in the complex and highly connected globalized world of trade, food security, and financial linkages.

Climate Scenario Storylines: The Dutch Climate Change Scenarios

National climate change programs are designed to provide an impact-oriented set of future climate conditions embedded in benchmark global climate scenario programs endorsed and reviewed by IPCC (RCPs, SSPs, CMIPs). Even for a given emission scenario or global warming level, an increasing uncertainty in global and regional climate change features remains present as the scenario horizon moves further into the future, and some form of selection or aggregation is needed. The Dutch climate change scenarios (KNMI'14, www.climatescenarios.nl) have carried out this aggregation at the national scale by condensing the available ~ 250 global and regional climate change projections into four discrete narratives, mutually discerned by choosing two elementary drivers of national climate and discerning contrasting values in the climate-enforced changes of these drivers: the global warming level and the regional response of atmospheric circulation have a large impact on hydroclimatic features in the region (Van den Hurk et al. 2014).

For each combination of elementary drivers, global and regional climate model simulations are collected and aggregated, to yield a comprehensive set of meteorological characteristics that have been adjusted to the needs of a wide range of sectoral stakeholders. Not only seasonally mean temperature and precipitation but also extreme values of daily and multi-day precipitation and snowmelt (to service flood risk practitioners), extreme winds (coastal surge), precipitation deficit aggregated to the growing season (agriculture), and extreme max and min daily temperatures (urban health).

Figure 4.1 gives a summary of the climate scenarios. It displays the essential decomposition of a large number of potential futures into four discrete scenarios and gives a brief narrative of the essential consequences of each of these scenarios for climate characteristics that are highly relevant for a wide range of stakeholders. The underlying storyline for each of the scenarios has a physical origin but is well understood by practitioners as a highly relevant source of uncertainty in climate response that adequately allowed formulating alternative policy scenarios for the low-lying Dutch Delta.

Skelton et al. (2017) evaluated the societal relevance and uptake capacity of different national climate scenario products, including the KNMI'14 scenarios. In their review, a strong interaction with users while scoping and constructing the climate scenarios is recommended. It is shown to contribute clearly to the credibility and legitimacy of the climate information, but challenges remain in making the scenarios relevant and useable for a different range of societal practitioners. For this, additional tailoring and user guidance are indispensable for efficient societal uptake of climate information (Berkhout et al. 2014).

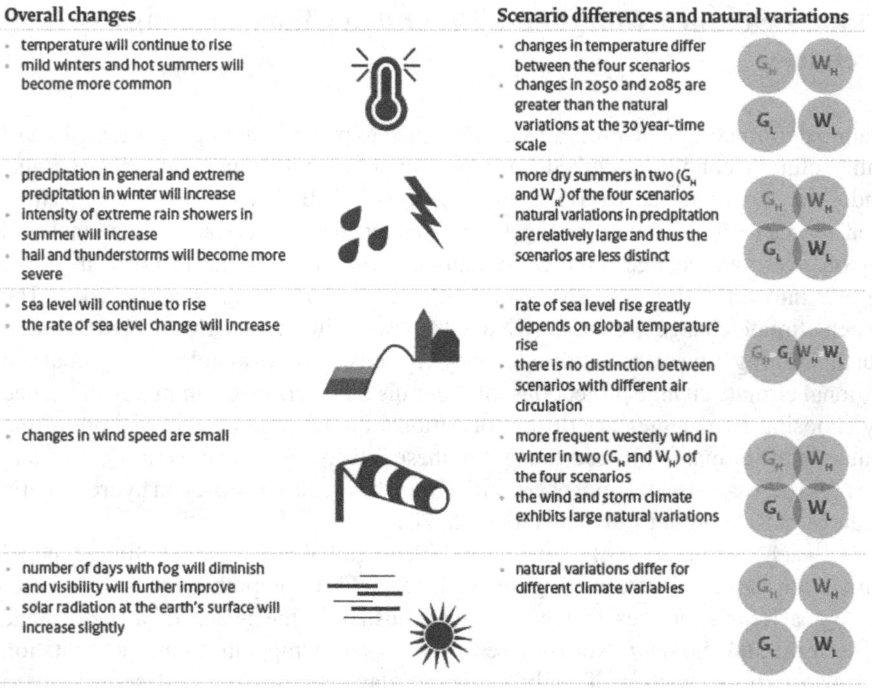
Overall changes
Scenario differences and natural variations
temperature will continue to rise
mild winters and hot summers will become more common
changes in temperature differ between the four scenarios
changes in 2050 and 2085 are greater than the natural variations at the 30 year-time scale
precipitation in general and extreme precipitation in winter will increase
intensity of extreme rain showers in summer will increase
hail and thunderstorms will become more severe
more dry summers in two (G_H and W_H) of the four scenarios
natural variations in precipitation are relatively large and thus the scenarios are less distinct
sea level will continue to rise
the rate of sea level change will increase
rate of sea level rise greatly depends on global temperature rise
there is no distinction between scenarios with different air circulation
changes in wind speed are small
more frequent westerly wind in winter in two (G_H and W_H) of the four scenarios
the wind and storm climate exhibits large natural variations
number of days with fog will diminish and visibility will further improve
solar radiation at the earth's surface will increase slightly
natural variations differ for different climate variables
G_H
W_H
G_L
W_L

Fig. 4.1 Summary of the overview of the KNMI'14 scenarios. Four scenario storylines are constructed varying in their level of global warming (*W* denoting high warming levels, *G* implying moderate warming) and the response of the regional atmospheric circulation (subscript *L* denoting a small change, *H* implying a large change). The driving conditions have large implications for regional climate characteristics that affect local societal impacts and their adaptation options, including many applications in water management, agriculture, and public safety (from www.climatescenarios.nl)

Climate Risk Storylines

COVID-19 convincingly demonstrates the difficulty to understand and foresee the complex cascades of shocks in our highly connected world. Although the parallels between COVID-19 and climate change impacts are only partially applicable, they clearly share the complexity of mapping consequences of remote disturbances on the European socio-economic risk profile. This complexity puts strong constraints on our ability to quantify this risk from a formalized probabilistic risk approach, by a combination of probabilities of (remote) hazards, exposure by means of socio-economic teleconnections (e.g. trade pathways), and vulnerability (European impact).

For a few years statements on the climate implications on isolated weather events are released by the application of so-called "attribution" studies, where the impact of climate change on the probability of the extreme event is quantified (Stott et al. 2016). The statements are strictly applicable to the characteristics of the event: any change in its appearance (time, location, drivers, impacts, etc.) will require a new "attribution" statement.

In the recent socio-economic history, a number of major climatic extreme events outside Europe have led to a noticeable impact on the European economy: US hurricanes affecting European (re)insurance and investment companies, strong and simultaneous adverse growing conditions in the world's "bread basket" regions, a flooding disrupting the supply of electronics, etc.

In an ongoing European H2020 research project RECEIPT (www.climatestorylines.eu), a number of event storylines, or climate risk storylines, are developed to map potential socio-economic consequences of extreme climate events outside Europe. The narratives are heavily inspired by experience from practitioners and stakeholders (Wilby and Dessai 2010), and new simulation and analysis techniques are developed to create analogs of these events for future climate conditions. These analyses do not aim to provide a comprehensive quantitative risk picture of any climate extreme in any region of the world but provide a strongly enriched picture of potential causal chains that may lead to (unexpected) impacts in downstream domains.

Conclusions and Recommendations

Underlying narratives are indispensable for the creation of credible, relevant, and legitimate climate information. Societal practitioners, the users of climate information, play a major role in defining the assumptions and contexts that need to be explored. Probabilistic approaches underlying many risk assessment methodologies are challenged when the context becomes very complex and stakes are high. To overcome some of these challenges, storyline approaches are maturing that enrich the picture of drivers, implications, and adaptation options of future climatic challenges. The definition of these storylines not only needs to comply with scientific standards to be credible and legitimate but also requires a thorough contextualization. This implies that a full understanding of all sources of uncertainty is not always achieved, but the inspiration provided by the climate storylines may make this uncertainty better conceivable and manageable.

References

Berkhout F, van den Hurk B, Bessembinder J, de Boer J, Bregman B, van Drunen M (2014) Framing climate uncertainty: using socio-economic and climate scenarios in assessing climate vulnerability and adaptation. Reg Environ Change 14(3):879–893

Eyring V, Bony S, Meehl GA, Senior CA, Stevens B, Stouffer RJ, Taylor KE (2016) Overview of the coupled model intercomparison project phase 6 (CMIP6) experimental design and organization. Geosci Model Dev 9:1937–1958. https://doi.org/10.5194/gmd-9-1937-2016

Riahi K, van Vuuren DP, Kriegler E, Edmonds J, O'Neill B, Fujimori S, Bauer N, Calvin K, Dellink R, Fricko O, Lutz W, Popp A, Cuaresma JC, Leimbach M, Kram T, Rao S, Emmerling J, Hasegawa

T, Havlik P, Humpenöder F, Aleluia Da Silva L, Smith S, Stehfest E, Bosetti V, Eom J, Gernaat D, Masui T, Rogelj J, Strefler J, Drouet L, Krey V, Luderer G, Harmsen M, Takahashi K, Wise M, Baumstark L, Doelman J, Kainuma M, Klimont Z, Marangoni G, Moss R, Lotze-Campen H, Obersteiner M, Tabeau A, Tavoni M (2017) The shared socioeconomic pathways and their energy, land use, and greenhouse gas emissions implications: an overview, global environmental change. https://doi.org/10.1016/j.gloenvcha.2016.05.009

Shepherd TG, Boyd E, Calel RA, Chapman SC, Dessai S, Dima-West IM, Fowler HJ, James R, Maraun D, Martius O, Senior CA, Sobel AH, Stainforth DA, Tett SFB, Trenberth KE, van den Hurk BJJM, Watkins NW, Wilby RL, Zenghelis D (2018) Storylines: an alternative approach to representing uncertainty in climate change. Clim Change 151:555–571. https://doi.org/10.1007/s10584-018-2317-9

Skelton M, Porter JJ, Dessai S et al (2017) The social and scientific values that shape national climate scenarios: a comparison of the Netherlands, Switzerland and the UK. Reg Environ Change 17:2325–2338. https://doi.org/10.1007/s10113-017-1155-z

Stott P et al (2016) (2016): Attribution of extreme weather and climate-related events. Wires Clim Change 7:23–41. https://doi.org/10.1002/wcc.380

Van den Hurk B, Jan G, van Oldenborgh G, Lenderink WH, Haarsma R, de Vries H (2014) Drivers of mean climate change around the Netherlands derived from CMIP5. Clim Dyn 42:1683–1697. https://doi.org/10.1007/s00382-013-1707-y

Van den Hurk BJJM, Hewitt C, Jacob D, Doblas-Reyes F, Döscher R, Bessembinder J (2018) The match between climate services demands and earth system models supplies. Clim Serv 12:59–63. https://doi.org/10.1016/j.cliser.2018.11.002

Vincent K, Daly M, Scannell C, Leathes B (2018) What can climate services learn from theory and practice of co-production? Clim Serv 12:48–58. https://doi.org/10.1016/j.cliser.2018.11.001

Wilby RL, Dessai S (2010) Robust adaptation to climate change. Weather 65:180–185

Zscheischler J, Westra S, van den Hurk B, Ward P, Pitman A, AghaKouchak A, Bresch DN, Leonard M, Wahl T, Zhang X, Seneviratne SI (2018) Future climate risk from compound events. Nat Clim Change 8:469–477

Chapter 5
On the Evaluation of Climate Change Impact Models for Adaptation Decisions

Thorsten Wagener

Abstract Detailed understanding of the potential local or regional implications of climate change is required to guide decision- and policy-makers when developing adaptation strategies and designing infrastructure solutions suitable for potential future conditions. Impact models that translate potential future climate conditions into variables of interest (such as drought or flood risk) are needed to create the required causal connection between climate and impact for scenario-based analyses. Recent studies suggest that the main strategy for the validation of such models (and hence the justification for their use) still heavily relies on the comparison with historical observations. In this short paper, the author suggests that such a comparison alone is insufficient and that global sensitivity analysis provides additional possibilities for model evaluation to ensure greater transparency and better robustness of model-based analyses. Global sensitivity analysis can be used to demonstrate that the parameters defining intervention options (such as land use choices) adequately control the model output (even under potential future conditions); it can be used to understand the robustness of model outputs to input uncertainties over different projection horizons, the relevance of model assumptions, and how modelled environmental processes change with climatic boundary conditions. Such additional model evaluation would strengthen the stakeholder confidence in model projections and therefore into the adaptation strategies derived with the help of these model outputs.

Keywords Impact models · Model validation · Uncertainty · Global sensitivity analysis · Stakeholder confidence

---

T. Wagener (✉)
Department of Civil Engineering, University of Bristol, Bristol, UK
e-mail: thorsten.wagener@uni-potsdam.de

Institute of Environmental Science and Geography, University of Potsdam, Potsdam, Germany

C. Kondrup et al. (eds.), *Climate Adaptation Modelling*, Springer Climate,
https://doi.org/10.1007/978-3-030-86211-4_5

Introduction

Human activity has become a geologic-scale force, changing landscape and climate at increasing rates in our effort to supply societies growing demand for water, energy and food. A fundamental scientific and societal question of our time is: how will water, energy and biogeochemical cycles be altered by this activity, and when and where will critical thresholds be reached (Gleeson et al. 2020)? This insight, if available at the local or regional scale, is needed to guide decision- and policy-makers in the development of adaptation strategies and in designing infrastructure solutions suitable for future conditions (Barron 2009).

To develop adaptation strategies, impact models are needed to translate climate signals into hydrological, ecological or other decision-relevant variables to understand potential implications of future climate for security issues related to water (droughts and floods), food, energy and health. An important question when using such impact models is how they have been evaluated regarding their ability to perform their task adequately (Wagener et al. 2010)? Most often, the task of addressing this question is referred to as model validation. In a recent review of the validation of resource management models for a wide range of uses including scenario modelling, Eker et al. (2018) found that data-based strategies to model validation still prevail. The use of historical observations, to show that a model can reproduce observed system responses, remains the main approach to demonstrate that a model is a valid representation of reality. The use of the term validation itself has been criticized, because it is overpromising in the sense that it suggests that a model has been established as being true, rather than adequate for the task at hand (Oreskes et al. 1994). Therefore the author uses the term evaluation in this short paper to suggest that we can ever only achieve an incomplete and conditional assessment of a model's suitability.

Any approach to evaluation suffers from multiple problems. First, the future might be significantly different from the past and demonstrating that a model is a realistic representation of the past system does not necessarily guarantee that it will reflect the future system. Many studies have therefore tried to create some type of resampling of the past to better reflect future conditions during model calibration/evaluation (e.g. Fowler et al. 2018; Singh et al. 2013), though this is only possible within limits. For example, even significant drought periods in the past will not fully reflect the combination of atmospheric, societal, land use and other conditions of the future. In some cases, modellers, therefore, prefer to run their models (especially across large domains) in uncalibrated mode, thus relying on the models' physical realism. This, however, regularly leaves significant performance gaps between models and observed behaviour—a discrepancy that is typically not propagated into the assessment of future model projections. Second, comparing the model to historical data can ignore the intended use of the model, which one might expect to be the main driver of the evaluation strategy. Klemeš (1986) in his seminal paper introduces multiple ideas for validation strategies in relation to intended model use, e.g. related to modelling land use change. These ideas are still rarely fully implemented. Focusing on fitting historical data is also emphasising model performance, rather than model robustness in the presence of unavoidable uncertainties. And third, comparing the model to historical data might also ignore the manner in which stakeholders gain trust in model predictions, especially related to modelling change (Eker et al. 2018). Stakeholders might, for example, care strongly whether the model structure reflects an understanding of the real-world system that is consistent with their own (Mahmoud et al. 2009).

In this brief paper, how the use of global sensitivity analysis can be beneficial in enabling model evaluation elements that are complementary to assessing the model fit to historical observations would be highlighted (Wagener and Pianosi 2019). It is important to stress that the comparison of historical data is not useful, rather that it is insufficient. So, in this paper, we will briefly discuss using three examples from previously published studies and conclude with some general remarks and suggestions.

Standard Evaluation Questions We Can (and Should) Ask Using Global Sensitivity Analysis

Do the Parameters That Reflect Possible Intervention Levers Adequately Control the Model Output?

A key role of impact models is to create causal links between cause and effect variables, especially in the context of developing adaptation strategies. We might, for example, want to understand how much land use choices such as deforestation/reforestation control the level of downstream flooding under future climate conditions, or we might want to know the level of influence of human activities such as groundwater pumping on the overall drought risk under potential future warming scenarios. We, therefore, have to demonstrate that the parameters reflecting these intervention levers (such as those describing land use or human activities) actually exert an adequate control on the model output consistent with our current understanding.

For example, Butler et al. (2014) performed a comprehensive variance-based sensitivity analysis of a doubled-CO_2 stabilization policy scenario generated by the globally aggregated dynamic integrated model of climate and the economy (DICE) (Fig. 5.1). The authors identified dominant processes by quantifying high sensitivities in model parameters relating to climate sensitivity, global participation in abatement and the cost of lower-emission energy sources. More importantly, in the context of this short paper, the authors did not find relevant sensitivities to other parameters such as those related to land use, that one might have expected to exert a stronger influence than the model shows. This result might suggest that certain intervention strategies cannot be assessed using the model in this particular example.

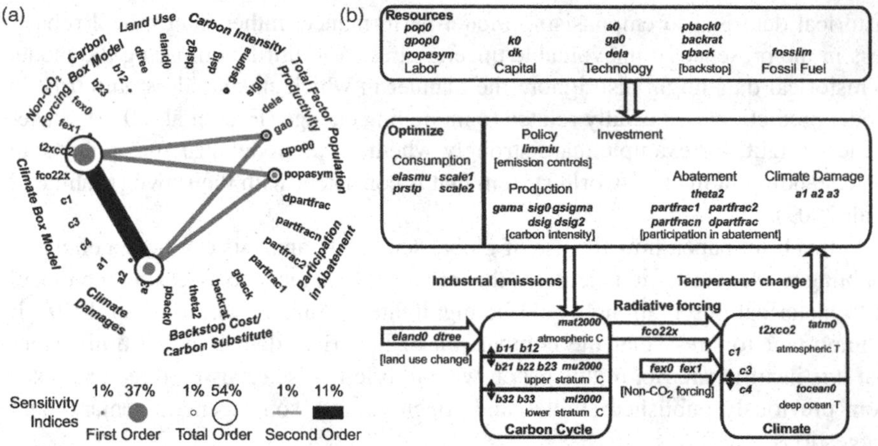

Fig. 5.1 Results of the study by Butler et al. (2014). **a** Sensitivities of the net present value of climate damages. The variance decomposition-based results are shown for first-order (filled circles), total-order (hollow rings), and second-order (connecting lines) indices. Diameters of the first- and total-order sensitivity circles are proportional to their respective sensitivity indices. Total sensitivities include first order and all higher-order (parameter interaction) sensitivities. The legend shows the extreme values for these metrics. Sensitivities of <1% are not shown; sensitivities of higher order than 2 are not explicitly shown. **b** Schematic diagram of the DICE model. Exogenous parameters are in italics. Parameters in bold blue italic are sampled in this study (Reproduced from Butler et al. 2014)

Are Dominant Uncertainties Changing Along the Projection Timeline?

It is further relevant to understand which uncertainties dominate the model output, especially over long time periods where levels of uncertainty might change considerably. Le Cozannet et al. (2015) used time-varying global sensitivity analysis to determine the factors that most strongly control the vulnerability of coastal flood defences over time (Fig. 5.2). They found that—for their question—global climate change scenarios only matter for long-term planning while local factors such as near-shore coastal bathymetry reflected in the wave setup parameter dominated in the short and mid-term (~over the next 50 years). The authors claim that wave setup uncertainty is often neglected in coastal hazard assessments studies. Global sensitivity analysis reveals that failing to incorporate the uncertainty in this process may invalidate conclusions and may lead to an overestimation of the effects of other drivers at least for short and mid-term planning periods. An assessment of the robustness of the model projections to input uncertainties thus has to consider the time-varying influence of these uncertainties.

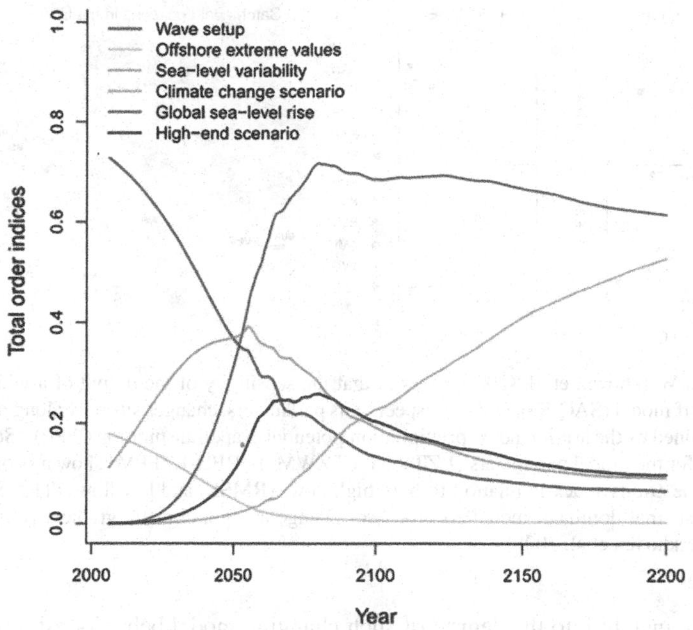

Fig. 5.2 The study by Le Cozannet et al. (2015) provides an example of using GSA to support long-term assessments; in this case of coastal defences. The figure shows the temporal sensitivity of predicted coastal defence vulnerability (specifically the output metric is the yearly probability of exceeding the threshold height of coastal defences). The figure shows that dominant drivers change significantly over time; for example global climate change scenario only matters beyond 2070 while offshore extreme values have no influence after that. Interestingly, for the time period up to 2050, the dominant factor is the 'wave set-up' parameter, which accounts for sea-level rise induced by wave breaking (Reproduced from Le Cozannet et al. 2015)

Are Dominant Modelled Processes Changing with Climate?

And finally, how strongly different modelled environmental processes control the output of adaptation models can vary strongly with climate or other boundary conditions. Models behave differently depending on the climatic boundary conditions they are applied in, regardless of the level of physics the model is based on (Rosero et al. 2010). Figure 5.3 shows some results of a study by van Werkhoven et al. (2008) who tested the sensitivity of a model's output (streamflow) to the parameters of a lumped rainfall-runoff model across 12 US catchments with very different climatic boundary conditions. The authors found (for both high flow and low flow conditions) that the controlling parameters varied considerably across climatic gradients. They further found that the spatial variability in sensitivity across catchments was similar to that observed within catchments when assessed across wet and dry years. This result suggests that for climate change projections, parameters (processes) that control the model behaviour for the historical period will likely differ from those that control the model output under new climatic boundary conditions. Global sensitivity analysis

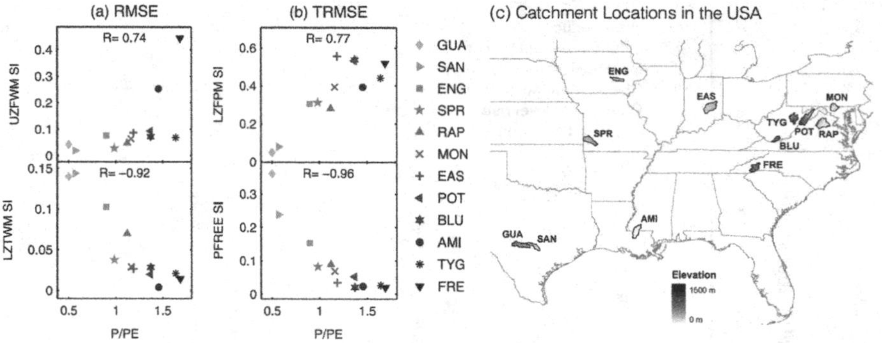

Fig. 5.3 van Werkhoven et al. (2008) showed that the sensitivity of the output of a widely used rainfall-runoff model (SAC-SMA) with respect to its parameters, changes strongly along a climatic gradient (defined by the aridity index: precipitation/potential evapotranspiration (P/PE)). Sensitivity indices (SI) for the model parameters (LZTWM, UFZWM, PFREE, LZFPM) show a strong correlation with the aridity index in relation to both high flows (RMSE) and low flows (TRMSE). This result suggests that dominant modelled processes change along a climatic gradient (Reproduced from van Werkhoven et al. 2008)

can provide insight into the degree of such changing model behaviour if a model is tested along climatic gradients.

Conclusions and Recommendations

Validation of impact models—an important task for developing adaptation strategies to climate change and for gaining stakeholder confidence—cannot be based on assessing a model's fit to historical data alone, even though such assessment can clearly play a role in establishing confidence in a model (Fowler et al. 2018; Eker et al. 2018). It is further important that evaluation strategies are linked to the intended model use (Klemes 1986). In this context, global sensitivity analysis is a valuable tool to complement any data- and performance-based validation strategy since it allows us to make the model and its simulations significantly more transparent. It is important to stress here that sensitivity analysis can be applied regardless of whether observations of the system response are available or not, thus making it very suitable for understanding the behaviour of models under potential future conditions (Wagener and Pianosi 2019). Possible questions for sensitivity analysis are:

Do the model parameters that are linked to potential adaptation strategies (e.g. via land use choices) exert expected levels of control on the modelled output?

Which uncertainties are likely to dominate the model output during the relevant assessment period?

Which modelled processes are likely to dominate the model output under the projected climatic conditions (rather than under the conditions for which historical observations are available) and are we confident in those estimated parameters?

Are model projections (and more importantly subsequent decisions) robust to input uncertainties?

How sensitive are the model projections to model assumptions?

Global sensitivity analysis can provide a valuable additional component to strengthen our confidence as well as the confidence of stakeholders in climate change impact models (Wagener and Pianosi 2019; Saltelli et al. 2020). Recent studies have further demonstrated that such sensitivity analysis can be performed even on highly complex models (Maples et al. 2020) or on those covering a global domain (Reinecke et al. 2019).

Acknowledgements This work on sensitivity analysis was initially supported by the UK Natural Environment Research Council (NERC) through the Consortium on Risk in the Environment: Diagnostics, Integration, Benchmarking, Learning and Elicitation (CREDIBLE) [NE/J017450/1]. Further support comes from a Royal Society Wolfson Research Merit Award and from an Alexander von Humboldt Professorship awarded to TW.

References

Barron EJ (2009) Beyond climate science. Science 326:643

Butler MP, Reed PM, Fisher-Vanden K, Keller K, Wagener T (2014) Inaction and climate stabilization uncertainties lead to severe economic risks. Clim Change 127:463–474. https://doi.org/10.1007/s10584-014-1283-0

Eker S, Rovenskaya E, Obersteiner M, Langan S (2018) Practice and perspectives in the validation of resource management models. Nat Commun 9:5359

Fowler K et al (2018) Simulating runoff under changing climatic conditions: a framework for model improvement. Water Resour Res 54(12):9812–9983

Gleeson T et al (2020) Illuminating water cycle modifications and Earth System resilience in the Anthropocene. Water Resour Res 56(4):e2019WR024957

Klemeš V (1986) Operational testing of hydrological simulation models. Hydrol Sci J 31(1):13–24

Le Cozannet G, Rohmer J, Cazenave A, Idier D, van de Wal R, de Winter R, Pedreros R, Balouin Y, Vinchon C, Oliveros C (2015) Evaluating uncertainties of future marine flooding occurrence as sealevel rises. Environ Model Softw 73:44–56

Mahmoud M et al (2009) A formal framework for scenario development to support environmental decision-making. Environ Model Softw 24:798–808

Maples S, Foglia L, Fogg GE, Maxwell RM (2020) Sensitivity of hydrologic and geologic parameters on recharge processes in a highly-heterogeneous, Semi-confined aquifer system. Hydrol Earth Syst Sci https://doi.org/10.5194/hess-24-2437-2020.

Oreskes N, Shrader-Frechette K, Belitz K (1994) Verification, validation, and confirmation of numerical models in the earth sciences. Science 263(5147):641–646

Reinecke R et al (2019) Spatially distributed sensitivity of simulated global groundwater heads and flows to hydraulic conductivity, groundwater recharge, and surface water body parameterization. Hydrol Earth Syst Sci 23(11)

Rosero E, Yang Z-L, Wagener T, Gulden LE, Yatheendradas S, Niu G-Y (2010) Quantifying parameter sensitivity, interaction and transferability in hydrologically enhanced versions of Noah-LSM over transition zones. J Geophys Res 115:D03106. https://doi.org/10.1029/2009JD012035

Saltelli et al (2020) Five ways to ensure that models serve society: a manifesto. Nature 582

Singh R, van Werkhoven K, Wagener T (2013) Climate change impacts in gauged and ungauged basins of the Olifants River, South Africa. Hydrol Sci J https://doi.org/10.1080/02626667.2013.819431

van Werkhoven K, Wagener T, Reed P, Tang Y (2008) Characterization of watershed model behavior across a hydroclimatic gradient. Water Resour Res 44:W01429. https://doi.org/10.1029/2007WR006271

Wagener T, Pianosi F (2019) What has global sensitivity analysis ever done for us? A systematic review to support scientific advancement and to inform policy-making in earth system modelling. Earth Sci Rev https://doi.org/10.1016/j.earscirev.2019.04.006

Wagener T et al (2010) The future of hydrology: an evolving science for a changing world. Water Resour Res 46:W05301

Chapter 6
Stress-Testing Adaptation Options

Robert L. Wilby

Abstract This technical contribution discusses ways of testing the performance of adaptation projects despite uncertainty about climate change. Robust decision making frameworks are recommended for evaluating project performance under a range of credible scenarios. Stress-testing options help to establish conditions under which there may be trade-offs between or even failure of project deliverables. Stress-tests may be undertaken for specified portfolios of management options, using models of the system being managed (including inputs and drivers of change), and then assessed against decision-relevant performance indicators with agreed options appraisal criteria. Field experiments and model simulations can be designed to test costs and benefits of adaptation measures. Simple rules may help to operationalize the findings of trials—such as 'plant 1 km of trees along a headwater stream to cool summer water temperatures by 1 °C'. However, insights gained from field-based adaptation stress-testing are limited by the conditions experienced during the observation period. These may not be severe enough to represent extreme weather in the future. Model simulations overcome this constraint by applying credible climate changes within the virtual worlds of system models. Nonetheless, care must be taken to select meaningful change metrics and to represent plausible changes in boundary conditions for climate and non-climate pressures. All stress-testing should be accompanied by monitoring, evaluation and learning to benchmark benefits and confirm that expected outcomes are achieved.

Keywords Climate change · Adaptation · Stress test · Weather generator · Field experiments

---

R. L. Wilby (✉)
Geography and Environment, Loughborough University, Loughborough L11 3TU, UK
e-mail: r.l.wilby@lboro.ac.uk

C. Kondrup et al. (eds.), *Climate Adaptation Modelling*, Springer Climate,
https://doi.org/10.1007/978-3-030-86211-4_6

Introduction

How can we be confident that investments in adaptation projects will deliver intended benefits *despite* deep uncertainty about climate variability and change?

This technical contribution discusses ways of evaluating the performance of adaptation measures. However, it is important to begin by acknowledging that climate change is not the only risk faced by human and natural systems. There are also concerns about resource depletion, biodiversity loss, environmental degradation, and especially human health. These are being driven by profound changes in demography, technology, global trade, public debt and urbanisation. Agencies are grappling with all these 'megatrends' whilst at the same time striving to meet policy goals around social cohesion, economic prosperity, national security and environmental sustainability. Nexus concepts are helpful in exposing the trade-offs that exist between climate change and connected policy areas. For instance, the climate-water-food-energy nexus should frame national and global efforts to achieve net zero emissions *whilst* simultaneously adapting to unavoidable climate change. Hence, this paper emphasizes the importance of adaptation planning that is integrated and mindful of multiple drivers of environmental change—we must avoid what some term 'climate exceptionalism' in our thinking.

Another important point of departure is to recognize that all 17 themes of the United Nations Sustainable Development Goals are intrinsically water-related. The hydrological community has traditionally been solution-orientated, but our generation faces perhaps the greatest array of water challenges in human history (Wilby 2019: 1464). Hence, this paper unashamedly views adaptation through a water lens, notwithstanding the above call for integrated planning and assessment. The next section describes a framework that emphasizes clarity about intended adaptation and/or development outcomes from the start. Two ways of evaluating attendant adaptation measures and investments are then discussed. By such 'stress-testing' we are seeking to better understand how various options might perform under credible climate *and* non-climate scenarios of system change. Fortunately, new tools and techniques are being developed to enable this via dedicated field experiments and systems modelling—each will be discussed in turn. Finally, some concluding remarks and practical recommendations will be offered.

Robustness and Resilience Frameworks

Conventional approaches to adaptation begin with climate model information as the basis for planning. Unfortunately, this 'predict-then-act' framework is soon confounded by growing uncertainties at each stage of the analytical chain: from the emissions scenario, to climate model selection, regional downscaling techniques, impacts modelling, ending with adaptation options appraisal (Wilby and Dessai

2010). Faced by wide ranges of uncertainty in outcomes, decision-makers may be forgiven for taking no action, delaying investments, or calling for further information.

A more fruitful strategy is to first accept that uncertainty is a fact of life—it may be better *characterized* by more research but is seldom *reduced* to a point where there is certainty about the consequences of a future action. By embracing the uncertainty, we are satisficing rather than optimising investments: some say we are seeking to minimize regret or maximize resilience through robust decision making (Weaver et al. 2013). Hence, robustness and resilience frameworks focus on testing decisions; climate model information is applied much later in the workflow to identify conditions under which there may be trade-offs or even break-points in performance. By concentrating on project goals and understanding key vulnerabilities, it is possible to target time and resources in more productive ways (ADB 2020). In situations where adaptation is a secondary objective there could be scope for light touch climate proofing (e.g., designs and materials for roads that will be upgraded every 10 years or so). Where addressing climate risks is the primary objective, or where there are long-lived investments, with risk of lock-in, high levels of precaution, or major economic consequences, detailed assessment is warranted (e.g., coastal defences to protect infrastructure from rising sea levels).

Robust decision making frameworks for adaptation option appraisal typically comprise of four main elements. These are as follows: (1) portfolios of management options; (2) models of the system being managed (including inputs and drivers of change); (3) project performance metrics; and (4) options appraisal criteria. Moreover, decision-centric frameworks are participatory and iterative in ways that enable managers and analysts to reach a shared understanding of key system vulnerabilities and adaptation goals (Fig. 1).

Let us imagine that authorities and private sector organisations have the legal power and/or responsibility for delivering a service—such as reliable water supplies—over a planning horizon that is potentially vulnerable to climate change. Ideally, these actors and their stakeholders will co-develop a portfolio of management options such as water saving measures, new or upgraded reservoirs and water

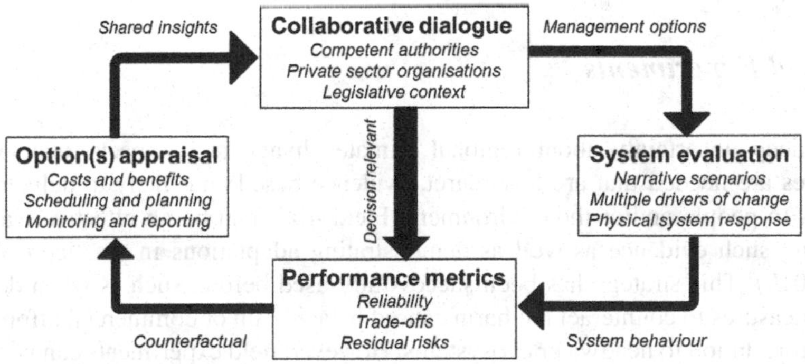

Fig. 1 An adaptation option appraisal framework. Adapted from Yates et al. (2015)

transfer schemes, source protection, environmental flows, artificial recharge, effluent reuse, water allocation and pricing controls.

Physical experiments or system models (see below) can then be used to evaluate (i.e., stress test) how such measures or combinations of measures perform under specified scenarios of change. These drivers may describe the future in narrative or numerical terms, but they must be credible and internally consistent. For instance, a hotter-drier climate change scenario might imply vegetation die-back, wildfires, or more dust on snow episodes that favour earlier snowmelt and more rapid rainfall-runoff. In other words, the stress-testing may need to be multi-dimensional to gain a more comprehensive view of direct and indirect climate risks, as well as non-climate threats (e.g., Ray et al. 2018).

Measures of system behaviour should also be meaningful to the decision-making context. Any trade-offs between outcomes should be apparent, such as reduced flood risk to property but less frequent or extensive rejuvenation of floodplain habitats (e.g., Poff et al. 2015). Moreover, it may not be economically or technically feasible to manage impacts from most extreme scenarios, so plans are needed for managing 'tolerable' risks against adaptation costs (e.g., Borgomeo et al. 2016). Ideally, multiple co-benefits will be measured too.

Finally, control experiments or counterfactual simulations are needed to benchmark outcomes 'with' versus 'without' adaptation. Options may be appraised using cost–benefit analysis and various adaptation pathways may be considered to schedule measures according to emergent climate and non-climatic pressures on the system. This presupposes a commitment to long-term monitoring of relevant drivers, well-defined trigger points for decisions, with monitoring of adaptation outcomes (e.g., Gell et al. 2019). Project goals or priorities will likely evolve, so the whole adaptation framework must be dynamic and open-ended. The following section gives more detail on the physical experiments and systems models than can be used to stress-test options.

Stress-Testing Methods

Physical Experiments

Given deep uncertainty about regional climate change and impacts, adaptation measures are needed that are low-regret, evidence-based and likely to deliver co-benefits to people and/or the environment. Field trials can be an effective way of obtaining such evidence as well as demonstrating adaptations in practice (Wilby et al. 2010). This strategy has been successfully used before, such as when developing measures to counteract the harm caused by acid rain or commercial afforestation/deforestation to headwater ecosystems. However, field experiments can be time and resource intensive, so they have to be carefully designed to test specific interventions—often using space-for-time substitutions to yield results quicker than the pace of climate change.

For example, the Loughborough University TEmperature Network (LUTEN) was established in 2011 to test a superficially straightforward adaptation measure: riparian shade management to 'keep rivers cool' (Johnson and Wilby 2015). In practice, the efficacy of shading rivers depends on a host of factors, not least the season, the location and area of any tree-planting along the river network, choice of species, their rates of growth, channel dimensions relative to tree height, and amount of local shading by the landscape.

Hence, a high-density network of paired air and water thermistors was installed in the Rivers Dove and Manifold, Midlands, UK, to gather data on space–time variations in these primary variables. The initial 36 test sites were chosen to represent a wide range of catchment, channel and bankside conditions, including open moorland, heavily wooded and deep Limestone gorge sections. Downstream water temperatures are further influenced by weirs, tributaries, ephemeral and perennial springs.

Long-term monitoring with modelling of shade revealed that approximately 1 km of riparian tree cover would lower daily maximum water temperatures by 1 °C in summer (Johnson and Wilby 2015). Moreover, the benefit of shade (relative to open reference sites) is greatest under hotter/drier/sunnier conditions. For instance, when air temperatures (Ta) are 25 °C, Tw can be ~3 °C cooler at sites with 77% compared with 43% upstream shade (Fig. 2). Such a thermal benefit might appear modest, but this could be the difference between lethal/sub-lethal conditions for biota during heatwaves.

The detailed field surveys further revealed significant local cooling by spring flows in middle and lower reaches of the rivers. Cool refugia like these should be carefully protected from non-climatic pressures such as trampling by cattle and fine sediments, as part of a broader programme of measures. Practicalities around land ownership, cost-benefits and maintenance of the riparian zone have to be resolved too. Nonetheless, simple rules of thumb like '1 km for 1 °C' help to operationalize the findings of the fieldwork.

Field experiments are ultimately constrained as a stress-testing tool by the range of weather conditions encountered during the period of observation (see: Wilby and Johnson 2020). Record homogeneity may also be affected by non-climatic changes.

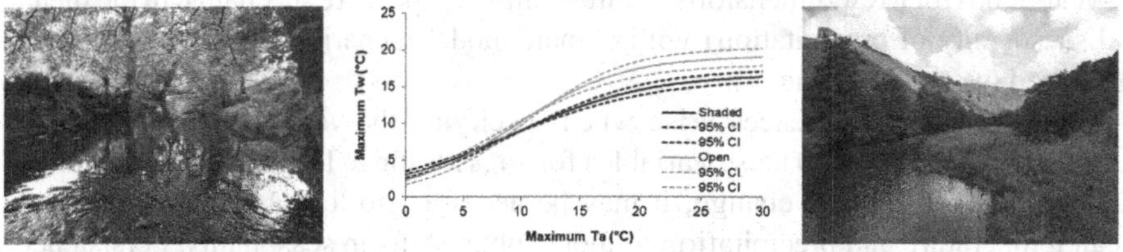

Fig. 2 Daily maximum water temperature (Tw) estimates for partially shaded (left) and open (right) sites.

For example, since the LUTEN monitoring began, there have been relatively few hot dry summers (until 2019) under which the thermal benefits of shade could be observed. Weir removals from mid-reaches of the drainage network had meanwhile impacted local river flow depths and velocities. Hybrid empirical-models (such as the logistic function in Fig. 2) can be fit to field data then used to extrapolate conditions at sites under climate change (such as higher Ta), but there are a host of associated stationarity assumptions. Alternatively, plot-scale or laboratory experiments (e.g., mesocosms) can apply changes in temperature, rainfall, water quality and even carbon dioxide concentrations under controlled conditions to assess outcomes with and without adaptations. However, these kinds of trial may be limited by the number of permutations of factors that can be practicably explored.

Systems Modelling

Systems modelling offers another means of evaluating adaptation options plus scope for more comprehensive, integrated assessment of risks. The technique involves running simulations with and without adaptations, given varied inputs representing the range expected boundary conditions. For example, Yates et al. (2015) took downscaled daily precipitation and temperature scenarios and then simulated the Denver Water, CO supply system using the Water Evaluation And Planning System (WEAP) with, and without, measures intended to protect reservoir storage during droughts. Hydrological model parameters were adjusted in line with climate scenarios to reflect potential changes in snowpack, land cover and soil properties. Accompany narratives described plausible drivers of the hydrology like 'fewer cold winters reduce mortality amongst infecting beetle populations' to adjust the vegetated area and evapotranspiration rate. With just three narrative scenarios it was shown that modest (but practically significant) adaptation benefits would be achieved.

Others implement more exhaustive stress-testing of adaptation measures, such as allowances (or headroom) for climate change in flood defence infrastructure (Broderick et al. 2019), portfolios of options to reduce the probability of water use restrictions (Borgomeo et al. 2016), or raising levees and changing reservoir operations to reduce flood damages and meet ecological objectives (Poff et al. 2015). Response surfaces are typically produced by simulating performance metrics (e.g., change in 20-year flood) for a few dimensions of future climate 'space' (e.g., change in the mean and seasonality of precipitation) with climate model scenarios overlain to indicate likelihood (as shown in Fig. 3).

Methodological differences arise when specifying the variable(s) and credible range(s) of adjustments to these variables for stress-testing. To really expose system vulnerabilities to climate change, it may be necessary to look beyond changes to mean temperature and precipitation to more subtle shifts in seasonality, persistence or extreme weather (e.g. Culley et al.2019). Plausible ranges for changes may be defined via stochastic weather generation of very large ensembles/rare events, or from the limits of widely adopted climate model ensembles (e.g. CMIP5 or CMIP6), via

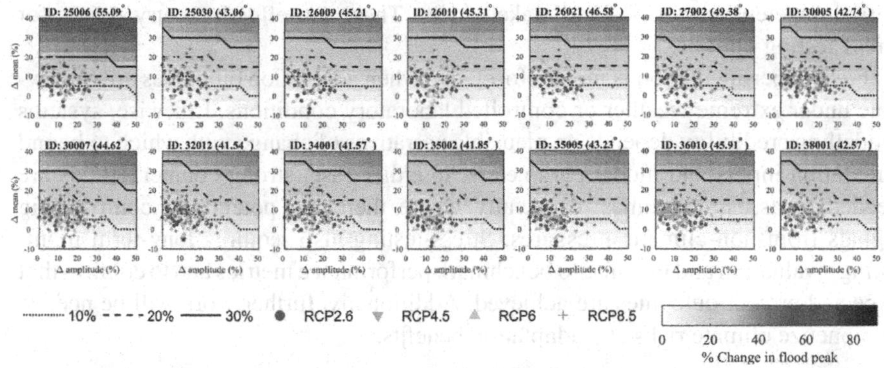

Fig. 3 Response surfaces for changes (%) in 20-year flood magnitude with incremental changes in the mean (x axis) and seasonality (y axis) of the annual precipitation cycle for selected catchments in Ireland based on CMIP5 models and four Representative Concentration Pathways. Source: Broderick et al. (2019).

expert panels and meta-analyses, or by a combination of approaches. For example, H++ scenarios for heat waves, droughts, floods, windstorms and cold snaps were developed from climate change scenarios at the margins or beyond the 10th to 90th percentile range of the 2009 UK Climate Change Projections (Wade et al. 2015).

High-end H++ scenarios of sea level rise were initially used to stress-test adaptation pathways for flood risk management within the Thames Estuary 2100 Plan. Now, credible maximum sea level change scenarios are informing the planning and testing of designs for new nuclear build in the UK. Under these highly precautionary circumstances, transparency about lines of evidence and working assumptions is essential; governance structures are also needed to ensure periodic review of evolving science, especially around key uncertainties about future meltwater contributions to sea level rise from the Antarctic and Greenland ice sheets.

Conclusions and Recommendations

This technical contribution asserts that deep uncertainty about the future climate and other major drivers of global change are not impediments to adaptation planning and options appraisal *provided*that decision-led (rather than scenario-led) frameworks are implemented. Stress-testing of physical (field-based) or virtual (model-based) systems under climate change, with and without adaptation, can reveal the efficacy of adaptations, as well as key system vulnerabilities and residual risks. Expert elicitation and ranking of options performance under different storylines may also be applied but there was insufficient space to discuss qualitative methods here (see for example, Brown et al.2015). Regardless of the approach taken, it is essential that portfolios of adaptation options and decision-relevant metrics are co-produced by analysts, with competent authorities and stakeholders. These are effective entry points for stress-testing options.

Field experimentation is most informative when adaptation outcomes are observable under extreme weather or controlled laboratory conditions. Likewise, systems modelling credibility depends on plausible narratives of change with which to bound simulation inputs and model parameters. In either case, strong counterfactuals are needed to assess outcomes, especially where there are potentially confounding signals from non-climatic pressures. Stress-testing also requires long-term monitoring, evaluation and learning to benchmark performance metrics and to confirm that expected project outcomes are achieved. Additionally, further work will be needed to monetize climate risks and adaptation benefits.

References

Asian Development Bank (ADB) (2020) Principles of climate risk management for climate proofing projects. ADB Sustainable Development Working Paper Series No.69. Metro Manila, Philippines

Borgomeo E, Mortazavi-Naeini M, Hall JW, O'Sullivan MJ, Watson T (2016) Trading-off tolerable risk with climate change adaptation costs in water supply systems. Water Resour Res 52:622–643

Broderick C, Murphy C, Wilby RL, Matthews T, Prudhomme C, Adamson M (2019) National assessment of climate change allowances for future flood risk using a scenario-neutral framework. Water Resour Res 55:1079–1104

Brown I, Berry P, Everard M, Firbank L, Harrison P, Lundy L, Quine C, Rowan J, Wade R, Watts K (2015) Identifying robust response options to manage environmental change using an ecosystem approach: a stress-testing case study for the UK. Environ Sci Policy 52:74–88

Culley S, Bennett B, Westra S, Maier HR (2019) Generating realistic perturbed hydrometeorological time series to inform scenario-neutral climate impact assessments. J Hydrol 576:111–122

Gell PA, Reid MA, Wilby RL (2019) Management pathways for the floodplain wetlands of the southern Murray Darling Basin: lessons from history. Rivers Res Appl 35:1291–1301

Johnson MF, Wilby RL (2015) Seeing the landscape from the trees: Metrics to guide riparian shade management in river catchments. Water Resour Res 51:3754–3769

Poff NL, Brown CM, Grantham TE, Matthews JH, Palmer MA, Spence CM, Wilby RL, Haasnoot M, Mendoza GF, Dominique KC, Baeza A (2015) Sustainable water management under future uncertainty with eco-engineering decision scaling. Nat Clim Chang 6:25–34

Ray PA, Bonzanigo L, Wi S, Yang YCE, Karki P, Garcia LE, Rodriguez DJ, Brown CM (2018) Multidimensional stress test for hydropower investments facing climate, geophysical and financial uncertainty. Glob Environ Chang 48:168–181

Wade S, Sanderson M, Golding N, Lowe J, Betts R, Reynard N, Kay A, Stewart L, Prudhomme C, Shaffrey L, Lloyd-Hughes B, Harvey B (2015) Developing H++ climate change scenarios for heat waves, droughts, floods, windstorms and cold snaps. Committee on Climate Change, London, p 145

Weaver CP, Lempert RJ, Brown C, Hall JA, Revell D, Sarewitz D (2013) Improving the contribution of climate model information to decision making: the value and demands of robust decision frameworks. Wiley Interdis Rev Clim Change 4(1):39–60

Wilby RL (2019) A global hydrology research agenda fit for the 2030s. Hydrol Res 50:1464–1480

Wilby RL, Dessai S (2010) Robust adaptation to climate change. Weather 65:180–185

Wilby RL, Johnson MF (2020) Climate variability and implications for keeping rivers cool in England. Clim Risk Manage 30:100259

Wilby RL, Orr H, Watts G, Battarbee RW, Berry PM, Chadd R, Dugdale SJ, Dunbar MJ, Elliott JA, Extence C, Hannah DM, Holmes N, Johnson AC, Knights B, Milner NJ, Ormerod SJ, Solomon D, Timlett R, Whitehead PJ, Wood PJ (2010) Evidence needed to manage freshwater ecosystems in a changing climate: turning adaptation principles into practice. Sci Total Environ 408:4150–4164

Yates D, Miller KA, Wilby RL, Kaatz L (2015) Decision-centric adaptation appraisal for water management across Colorado's Continental Divide. Clim Risk Manag 10:35–50

Chapter 7
Reducing and Managing Uncertainty of Adaptation Recommendations to Increase user's Uptake

Margarita Ruiz-Ramos and Alfredo Rodríguez

Abstract There are many challenges that adaptation science faces for an effective application of the results and recommendations found. Among the most important are the estimation, management and interpretation of uncertainty. In this article, we present our approach to managing uncertainty in agricultural projections using a combination of techniques to identify uncertainties, exclude unviable outcomes, consider possible futures probabilistically, and select the most robust projections for adaptation. Through an example of the adaptation of winter wheat in Spain, we show how this approach is effective in increasing the probability of avoiding maladaptation and improving the applicability and assimilation of scientific results by users.

Keywords Adaptation response surface, Ensemble Outcome Agreement, adaptation confidence

Introduction

According to the 2019 report of the World Economic Forum (WEF 2019), extreme weather events and failure on climate-change mitigation and adaptation are the two risks with the highest probability and the highest impact among the analysed in the report, with other issues such as water crises following them closely. These risks, when they turn into events, impose severe losses and damages to both private and public goods and activities, from small producers to insurance companies and administrations.

During the latest decades, a number of adaptation studies have been conducted. However, there are many challenges that adaptation science has to face for an effective application of the results and recommendations found. Among the most important

M. Ruiz-Ramos (✉)
CEIGRAM, Universidad Politécnica de Madrid, 28040 Madrid, Spain
e-mail: margarita.ruiz.ramos@upm.es

A. Rodríguez
Department of Economic Analysis and Finances, Universidad de Castilla-La Mancha, 45071 Toledo, Spain
e-mail: alfredo.rodriguez@uclm.es

C. Kondrup et al. (eds.), *Climate Adaptation Modelling*, Springer Climate,
https://doi.org/10.1007/978-3-030-86211-4_7

are model errors, scarcity of observation records, lack of knowledge on the trade-offs with mitigation, political, social and behavioural barriers, and uncertainty estimation, management, and interpretation. This paper is focused on this last issue related to the adaptation of the agricultural sector to climate change.

Uncertainty of the adaptation assessments has many sources: The main one is the degree of success of mitigation, i.e., the scenario or RCP, which will determine the level of warming, seasonal patterns, and extremes with which adaptation will have to deal (Dosio and Fischer 2018). Another relevant source of uncertainty is model flaws and small observational records, especially for not mainstream measurements. This is valid for climate and impacts: for instance, in the climatology of extremes, or for crops other than the most common staple ones (wheat, maize, etc.), data and knowledge are poorer, as it is modelling performance consequently. Therefore there is always a degree of model error (that we can know when comparing to observations) and model uncertainty (e.g., in the case of future projections for which we cannot know in absence of a comparison reference). Efforts to reduce it are needed and approaches to do so are being developed and put in place, but also an improvement of its understanding and communication will help to handle the remaining uncertainty once such approaches have been applied. In this paper, we present our approach to handling the uncertainty of agricultural projections.

Material and Methods

Climate Data

Observed data for Spain were taken from a station of the Spanish Meteorological Agency (AEMET) located al Lleida (NE Spain) during the period 1980–2010. Variables at a daily scale were minimum and maximum temperatures (Tmin and Tmax, respectively), precipitation (P), solar radiation or sun exposure (as a proxy for calculating solar radiation), humidity and wind speed. These data were used for crop model calibration and as baseline weather. P and temperature (T) baseline values were perturbed, to conduct a sensitivity analysis, by using a "change factor" approach in combination with a seasonal pattern of the T and P changes (Fronzek et al. 2010).

As for climate change projections, RCMs simulation outputs from the Euro-CORDEX domain (www.euro-cordex.net, last access: 7 July 2020) are used. These simulation outputs, with much higher resolution than GCMs, are particularly interesting for adaptation and mitigation studies. In this case, the 0.11° resolution was selected. The simulations are forced by different GCMs and emission pathways (RCP2.6, RCP4.5 and RCP8.5, van Vuuren et al. (2011)).

Crop Data and Models

Crop and soil data were required to calibrate crop models. Data were retrieved from published experiments from our research group or from other collaborating groups and literature. These data consisted of descriptions of the genetics of the crop and the cultivar in terms of phenology and growth potential. Also, information about the soil and common management and crop responses to favourable and stressed situations was required.

CERES-wheat crop model within DSSAT platform (Hoogenboom et al. 2019) was used for simulating wheat responses to climate change.

Adaptation Modelling

Adaptation consisted of a combination of changes of sowing dates, cultivar and water management. Baseline simulation referred to a winter cultivar describing those currently sown in the region, at the beginning of November, under rainfed conditions. Cultivar features modified were related to phenological characteristics, the vernalisation requirements and the length of the phenological phases. In the study region spring wheat can be sown in the same sowing dates as winter wheat; the decision on autumn rain. Shorter and longer growth cycles were simulated (+-20% of season length), and the sowing date was advanced and delayed up to 45 days. Another option was supplementary irrigation, consisting of a single application of 40 mm at flowering. These adaptations were simulated separately and combined, by an ensemble of 17 members made up of 14 crop models and 17 modelling groups from MACSUR project (Ruiz-Ramos et al. 2018).

Approaches for Uncertainty Reduction

Several approaches have been applied depending on the estimate of the initial uncertainty. These are the bias adjustment (always applied), the ensemble modelling, the AOCK approach, the response surfaces and the Ensemble Outcome Agreement (EOA) index.

Ensemble modelling is a technique widely used in climate and also in agricultural projections, ideally, combining outputs from several climate models and crop models allow estimating the joint uncertainty from climate and crop modelling. Application of the AOCK approach, which consists in defining study-specific disqualifying criteria to be part of the ensemble, prevents for including inconsistencies that affect ensemble results (e.g., rainfed simulation projecting more yield than irrigated simulations in a dry environment). This approach is able to provide insights and improve ensemble effectiveness. Nevertheless, this has to be combined with other strategies to assure ensemble quality as the ensemble composition representing the spread of model projections and the use of medians instead of the mean as an average descriptor.

For situations of high uncertainty, the response surfaces approach is used. An Impact Response Surface (IRS) consists of a plotted surface that depicts the response of a studied variable (the so-called impact variable, e.g., crop yield) to combined changes in two explanatory variables (e.g., P and T). An Adaptation Response Surface (ARS) plots the difference between the impact variable responses (e.g., yield) with and without adaptation being considered, usually as a percentage of change. This metric is defined as the "adaptation value". It measures the effect of adaptation under a given combination of perturbations of T, P and [CO_2] compared to the not adapted situation under the same perturbations. Another metric called "recovery value" refers to the difference between the yield response including an adaptation option and the baseline yield response (i.e., for an unperturbed simulation, 360 ppm of [CO_2] and un-adapted management). The "recovery value" measures the capacity of an adaptation option to maintain the yields of the baseline simulation under unperturbed conditions.

Finally, we develop an index of wide applicability for reducing the uncertainty, the so-called Ensemble Outcome Agreement (EOA) to assessing the confidence of the decisions taken regarding a threshold level, such as recommended adaptation options that are projected to provide positive adaptation values by the ensemble. Specifically, this index measures how the effect of changes in composition and size of a multi-model ensemble (MME) to evaluate the level of agreement between MME outcomes with respect to a given hypothesis (e.g., that adaptation measures result in positive crop responses). Definition of the index can be found in Rodríguez et al. (2019).

Results and Discussion

As application of the multi-model ensemble modelling, AOCK and IRS and ARS, recommendations for adaptation of wheat in Spain were produced as shown in Table 3 of Ruiz-Ramos et al. (2018). When EOA was applied to this dataset, Fig. 3 of Rodríguez et al. (2019) was obtained. The main added value of this further improvement was that by assigning an EOA value to every adaptation option, those that are otherwise promising in terms of adaptation response but show low confidence (i.e., low values of EOA) can be discarded, in favour of those with more confidence. Revisiting the recommendations of Ruiz-Ramos et al. (2018) in light of the EOA index generally resulted in an improvement by narrowing the range for which the adaptation options were effective.

While most of the winter-wheat based adaptations under rainfed conditions provided a very low value of EOA, at least 1 adaptation option, mainly based on spring wheat, was found with a high EOA value for every perturbation combination of P and T. This supports the idea that adaptation would be possible under a wide range of future conditions. The number of effective adaptation options highly increased when supplementary irrigation was considered; in this case, some options provided positive results also for winter wheat.

As concerning sowing dates, EOA analysis supports the recommendations done Ruiz-Ramos et al. (2018) for adaptation, while for recovery the main difference was lower confidence reported by EOA for many cases. For adaptation response, recommendations for standard and longer cultivars were confirmed with very high or maximum confidence, while the confidence level was variable for high perturbations for rainfed spring wheat. For recovery response, the perturbation range for which spring wheat-based options were effective was smaller than previously estimated. For both adaptation and recovery response, the confidence level for shorter cultivars of both winter and spring wheat was lower than for the other cultivars, including some adaptation options previously recommended that now should be excluded. As a consequence, the revised recommendations would be to focus on early and standard sowing dates combined with standard and longer cultivars for meeting both adaptation and recovery targets under moderate perturbations with very high confidence, while there would be chances of achieving only adaptation benefit (impact reduction) with these options for severe perturbations. When the aim would be just to adapt, short spring cultivars could also be used with high confidence. The study demonstrates how omitting this analysis would result at least in a number of misleading recommendations under certain perturbations, resulting in maladaptation.

Conclusions and Recommendations

Adaptation of winter wheat in Spain relies on finding local optimal combinations of cultivars and management. In NE Spain, winter wheat under future conditions would benefit from switching to spring cultivars (i.e., no vernalisation requirements) with standard and longer cycle length, combined with early and standard sowing dates in autumn.

Techniques for managing uncertainty of agricultural projections or other projections depending on an uncertain future climate, such as EOA and others are as necessary as effective to increase the probability to avoid maladaptation and enhance the applicability and uptake of scientific results by users.

Acknowledgements Spanish National Institute for Agricultural and Food Research and Technology and Agencia Estatal de Investigación grant MACSUR02-APCIN2016-0005-00-00 and by the Comunidad de Madrid (Spain) and Structural Funds 2014-2020 (ERDF and ESF), project AGRISOST-CM S2018/BAA-4330.

References

Dosio A, Fischer EM (2018) Will half a degree make a difference? Robust projections of indices of mean and extreme climate in Europe Under 1.5°C, 2°C, and 3°C global warming. Geophys Res Lett 45(2):935–944

Fronzek S, Carter TR, Raisanen J, Ruokolainen L, Luoto M (2010) Applying probabilistic projections of climate change with impact models: a case study for subarctic palsa mires in Fennoscandia. Clim Change 99(3–4):515–534

Hoogenboom G, Porter CH, Shelia V, Boote KJ, Singh U, White JW, Hunt LA, Ogoshi R, Lizaso JI, Koo J, Asseng S, Singels A, Moreno LP, Jones JW (2019) Decision support system for agrotechnology transfer (DSSAT) version 4.7.5. DSSAT Foundation, Gainesville, Florida, USA. https://DSSAT.net

Piani C, Haerter JO, Coppola E (2010) Statistical bias correction for daily precipitation in regional climate models over Europe. Theoret Appl Climatol 99(1–2):187–192

Richardson EA, Seeley SD, Walker DR (1974) A model for estimating the completion of rest for "Redhaven" and "Elberta" peach trees. HortScience 1:331–332

Rodríguez A et al (2019) Implications of crop model ensemble size and composition for estimates of adaptation effects and agreement of recommendations. Agric Meteorol 264:351–362

Ruiz-Ramos M et al (2018) Adaptation response surfaces for managing wheat under perturbed climate and CO2 in a Mediterranean environment. Agric Syst 159:260–274

Ruiz-Ramos M et al (2016) Comparing correction methods of RCM outputs for improving crop impact projections in the Iberian Peninsula for 21st century. Clim Change 134(1):283–297

Shaltout AD, Unrath CR (1983) Rest completion prediction model for "Starkrimson Delicious" apples. J Amer Soc Hort Sci 108(6):957–961

van Vuuren DP et al (2011) The representative concentration pathways: an overview. Clim Change 109:5–31

WEF (2019) The GlobalCompetitiveness Report. World Economic Forum, Geneva, p 666

Part II
Hazard, Exposure and Vulnerability Modelling

Ad Jeuken
Deltares, Delft, The Netherlands
ad.jeuken@deltares.nl

Introduction

Hazard, exposure and vulnerability are the three main determinants for risk. Though stemming from the disaster risk management community, this risk concept is now widely applied to climate change impact and adaptation analysis and further highlighted in the IPCC AR5 report, see Fig. 1.

For (the modelling of) adaptation and climate change impacts the decomposition of risks into different components offers a helpful framework for both understanding

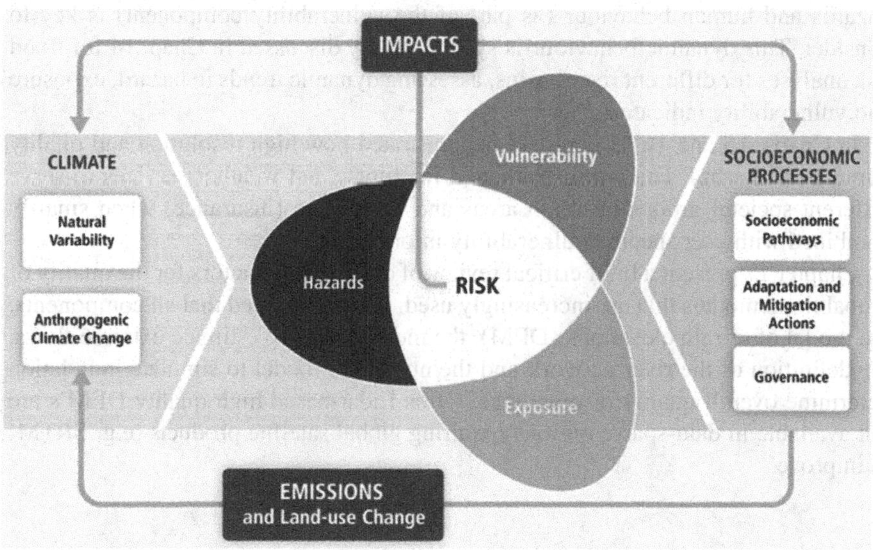

Fig. 1 Components of climate risks and its key drivers

the main causes and guiding of action to tackle them. An example of how different adaptation options can be explored under three categories, hazard-, exposure- and vulnerability reducing measures, and what this requires in terms of modelling efforts, is presented in Chap. 10 for coastal systems.

A main challenge is the representation of climate uncertainty in the analysis. Traditionally a selection of down-scaled GCM projections is used to represent a plausible input to hazard models. An emerging group of scholars is applying novel bottom-up approaches that apply stress tests to analyse plausible combinations of stressors that lead to the highest vulnerability, potentially surpassing critical system thresholds and leading to tipping point behaviour. In Chap. 11 the use is demonstrated of impact response surfaces to depict the modelled sensitivity of indicators to climate and socioeconomic drivers across a plausible range of perturbations, accounting for uncertainty where applicable. This allows the decision-maker to assess the evolving likelihood of exceeding stakeholder-defined thresholds of impact and hence informs the urgency to act under certain evolution of climate change. A similar approach for stress testing adaptation options is presented in Chap. 7 by R.L. Wilby.

In Chap. 12 the need for better quantification of the hazard reducing effect of nature-based solutions, in this case the mitigation of land slide risks by vegetation, is stressed. As vegetation behaviour is dynamic in time and space, the effects on slope stability are difficult to model. Another example of challenges that are associated with biotic system components in modelling is presented in Chap. 13. Here it is illustrated how to model the sensitivity of crops changes in response to climate and soil-hydrology and how crop rotation as an adaptation option requires insight from modelling on interactions (carry over effect) over time from one rotation to another.

As there are important dynamic interactions between abiotic and biotic components to consider in adaptation modelling, also the dynamic interaction between hazards and human behaviour (as part of the vulnerability component) is key to consider. This dynamic behaviour is systematically discussed in Chap. 14 for flood risk analyses for different river basins, assessing dynamic trends in hazard, exposure and vulnerability indicators.

In Chaps. 15 and 16 its is nicely demonstrated how high resolution and quality climate and hazard data are instrumental in climate and weather services to serve different societal groups in urban areas and agriculture (insurance) when smartly combined with user specific vulnerability information.

Chapter 17 present a brief critical review of determining factors for the quality of global flood models that are increasingly used. It is highlighted that all components, the model of terrain elevations (DEM), the method used to estimate extreme flows, the definition of the river network and the numerical model to simulate inundation determine overall quality of outcomes. Often Lidar-based high quality DEM's are not available in data-sparse regions requiring global satellite products (e.g. SRTM) to improve.

Chapter 8
Modelling Risk Reduction Measures to Minimise Future Impacts of Storms at Coastal Areas

Óscar Ferreira

Abstract Coastal storms often cause damages and losses in occupied areas. Under climate change conditions (i.e. sea-level rise and increased frequency of extreme sea levels) and increasing human occupation, the consequences of coastal storms will be amplified if no adaptation actions are implemented. The selection of the best possible coastal management measures to reduce risks at coastal areas, considering costs, effectiveness and acceptance, will be mandatory in the future. This work presents a generic approach to model disaster risk reduction measures at coastal areas, including climate change effects. The proposed methodology is adaptable to any coastal region and can be used to test (and improve) management options at a broad number of coastal areas. It can also be used to define a timeframe for the implementation of the defined measures since not all risk reduction measures, under a climate change scenario, need to be implemented at the same time. This would help to optimise implementation costs while reducing the risk to the occupation and people.

Keywords Coastal hazards · Climate change · Preparedness · Coastal management

Introduction

Storms impacting sandy coastal areas produce hazards such as erosion and inundation that, in turn, promote risk to life and property damage in occupied areas, and the alteration and/or fragmentation of habitats (Ferreira et al. 2017). Coastal damages and risks are expected to increase in the near future not only in association with climate change (e.g. sea-level rise (SLR), change in frequency and magnitude of storms) but also due to increasing human occupation in coastal areas (e.g. van Dongeren et al. 2018). In the next decades, SLR will likely become the dominant driver in erosion/flooding risk at coastal areas and may escalate that risk by up to 300% over the next 3 decades (Wahl and Plant 2015). Because of climate change, Vousdoukas et al. (2018a) project an increase of the global average 100-year extreme sea levels

Ó. Ferreira (✉)
Universidade do Algarve, CIMA, FCT, Faro, Portugal
e-mail: oferreir@ualg.pt

C. Kondrup et al. (eds.), *Climate Adaptation Modelling*, Springer Climate,
https://doi.org/10.1007/978-3-030-86211-4_8

between 2000 and 2100 of 58–172 cm. Considering existing trends in shoreline dynamics, combined with coastal recession driven by sea-level rise, Vousdoukas et al. (2020a) estimate a total retreat exceeding 100 m for almost half of the world's sandy beaches, by the end of the century.

In Europe, the expected annual damage due to coastal floods is expected to increase by three orders of magnitude by the end of the century if no adaptation is taken (Vousdoukas et al. 2018b). By the 2080s, between 13 and 123 million additional people will face annual coastal floods worldwide, assuming no upgrade in adaptation measures for an additional SLR between 0.19 m and 0.68 m (Brown et al. 2019). By 2100, without adaptation, 0.2–4.6% of the global population is expected to be flooded annually under 25–123 cm of global mean SLR, with expected annual losses of 0.3–9.3% of the global gross domestic product (Hinkel et al. 2014).

With this view of the future, coastal authorities need to assess the level of impact and the risk in their coastal zones and implement tested and feasible Disaster Risk Reduction (DRR) and PMP (prevention, mitigation and preparedness) measures (van Dongeren et al. 2018). Considering that a substantial proportion of the threatened coastal areas are in densely populated areas, there is a need for the design and implementation of effective adaptive measures (Vousdoukas et al. 2020a). For that purpose it is paramount to first test and validate those measures, analysing their effectiveness and choosing the optimal ones (higher effectiveness for lower cost). That can be done by implementing such measures at existing models and simulating their response and behaviour to coastal storms under climate change scenarios. This paper presents and discusses models and approaches to test DRR measures and their effectiveness in minimising the impacts of coastal storms at occupied areas.

Modelling the Impact of Coastal Storms

Coastal retreat and flooding can be estimated by using simple formulations or models. For overwash assessment, the most commonly used method is to estimate runup using empirical formulations that require offshore wave conditions and beach slope. The computed runup can be added to a given sea level and compared with the existing morphology to estimate the overwash potential, depth and/or extension. For coastal inundation, bathtub approaches are still often used, simply comparing the estimated sea level (for a given return period) with the morphology or occupation elevation. The storm-induced retreat/erosion can also be computed by using relatively simple analytical models (e.g. Larson et al. 2004; Mendoza and Jiménez 2006). These models use relatively simple formulations that integrate driving mechanisms (such as wave height, storm duration and sea level) jointly with the morphological and sedimentological characteristics of the coastal area (e.g. dune height, berm width, beach slope or grain size). Such simple methods have, however, several limitations, mainly at complex environments, such as engineered coastlines or areas with complex geomorphology. Furthermore, the hazard and consequent risk can change due to feedback mechanisms. The lowering of a dune by overwash will lead to an increase in the hazard when compared with the initial situation/morphology. These feedback mechanisms can also occur differently alongshore, as a function of morphological variability. In cases where feedback mechanisms may be highly relevant, simple models and formulations may not fully reflect the impacts associated with a given event. In those cases, process-based models with high-resolution topo-bathymetric grids, after validation and calibration, may be helpful to better understand the hazard in coastal areas (Ferreira et al. 2017).

Process-based models, like XBeach (Roelvink et al. 2009), can be employed to determine both coastal erosion and flooding, for all regimes and with great detail but requiring a higher level of computational complexity and available data for model calibration. Process-based models reproduce the processes occurring on coastal areas during a storm, containing the essential physics of dune erosion, overwash, avalanching, swash, infragravity waves and wave groups. Inundation models, such as LISFLOOD (Bates and De Roo et al. 2000), which account for lateral connectivity and permeability, can also be used to better represent the inundation area. Thus, the recommended models to determine the hazard associated with episodic erosion and/or flooding are open-source process-based nearshore storm impact models such us XBeach (for erosion and overwash) or XBeach coupled with the LISFLOOD (for marine flooding). These models have been applied to a vast number of diverse coastal areas and have extensive use and validation, providing confidence in their application (Ferreira et al. 2018).

Process-based models make use of complex-modelling techniques and require a high amount of data. They can be used on a stand-alone basis, but are normally integrated within a more complex processing scheme that includes data import from external sources, data processing, perform the runs, data post-processing and simplification and final results exportation. For the mentioned scheme, model trains are often developed (see example at Ferreira et al. 2018), starting from the incorporation of available data from other operational systems and downscaling storm conditions to local hazards. They should take into consideration, on their implementation, the availability of suitable regional data sources or forecast systems, the dominant physical and morphological conditions that control the storm processes, the selected onshore hazards to be modelled, and the selected receptors to be impacted. In the case of modelling climate change, they must also take into consideration the existence of adequate regional or local climate change predictions to be incorporated as input values. The results of the high-resolution hazard models are translated into impact using damage curves or any other relationship that relates hazard with the damage at the receptors.

Modelling Risk Reduction Measures Including Climate Change

When considering the impact of climate change at coastal areas, most of the current approaches assume that the existing morphology and occupation will be maintained and only the forcing conditions will change (e.g. Vousdoukas et al. 2018a; Brown et al. 2019). Recent works often use simple approaches to adaptation (e.g. vertically rising dykes, Vousdoukas et al. 2020b) and do not run computational demanding morphodynamic models. Only rarely DRR measures are fully modelled and tested (e.g. Plomaritis et al. 2018; Ferreira et al. 2019) and even less frequently they incorporate climate change effects.

To support and aid to achieve the most informed and best possible decisions regarding risk prevention, models can be used to simulate and evaluate the effectiveness of current and potential future DRR measures under storm conditions and even climate change. This is achieved by simulating historical and climate change-related storm scenarios with and without DRR measures in place (Fig. 8.1). From these simulations, it is possible to obtain predictions of local flooding and erosion, which combined with characteristics of the local population, built environment and infrastructure, allow to compute storm impact and the effectiveness of the measures. Models can also be used to evaluate how effective a DRR measure or a combination of measures will be on the reduction of the impact of storm events. Those measures can be split into different types that require different modelling approaches, as explained by Jäger et al. (2018). "Exposure-reducing measures" move receptors out of high-risk areas by temporarily evacuating people or permanently relocating residential areas. For instance, house removal (or relocation), will remove the exposed elements (reducing risks) while the computed hazard may remain the same. For these measures, receptors are removed from the model, which do not need to be rerun (Fig. 8.1), since the hazard level will be the same, for current conditions, while the risk is reduced due to the removal of the occupation. "Pathway-obstructing measures" change the morphology and hence its interactions with waves and water levels. That is the case of beach nourishment or dune recovery that obliges to introduce a new morphology and to rerun the models for all desired conditions (Fig. 8.1) since the hazard level

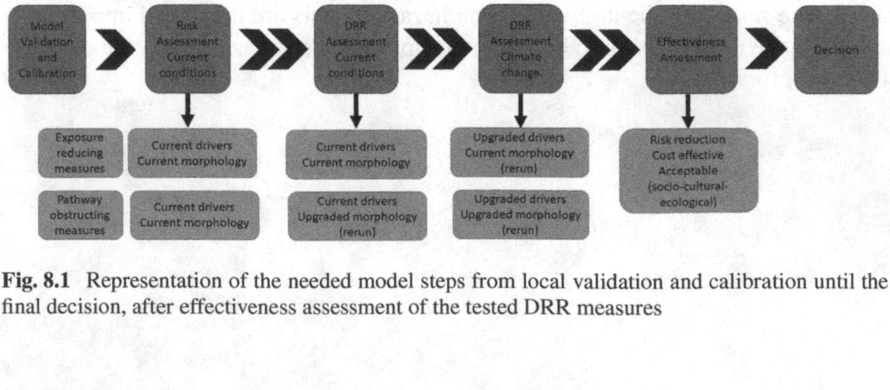

Fig. 8.1 Representation of the needed model steps from local validation and calibration until the final decision, after effectiveness assessment of the tested DRR measures

will not be maintained. The third type of DRR measures is "vulnerability-reducing measures", which include flood protection for individual receptors (e.g. sandbags) or raising awareness at the population. This type of DRR measures is not easy to model, although there are attempts to assess their effectiveness (see Cumiskey et al. 2018).

The impact of predicted future climate scenarios (e.g. sea-level rise and extreme sea levels), based on available projections at the regional or local scale, under the Representative Concentration Pathway 8.5 or other adequate estimates, can be incorporated in the models to assess the future effectiveness of DRR measures. In such cases, new input variables (e.g. wave height and total sea-level for a given return period or a set of forecasted conditions for a given year in the future) will be required, depending on the chosen scenario and year (e.g. 2050, 2100), obliging to rerun the model for each defined condition (Fig. 8.1). Thus, for DRR assessment under climate change scenarios, both types of measures (exposure-reducing and pathway-obstructing) most probably imply the model's rerun.

The choice of the DRR measures to be tested should be done by expert judgment in consultation with end-users and stakeholders to ensure their future integration into management plans and to guarantee that informed and scientific-based coastal management decisions are taken. The final decision, after testing (Fig. 8.1), should consider the measured effectiveness, the social-cultural and the ecological acceptance of the measures, to ensure their sustainability and approval. For the effectiveness index (Ie) Ferreira et al. (2019) proposed the use of Eq. 8.1 for each simulated DRR:

$$Ie = 100\%x \frac{(\%damagecurrentsituation - \%damagedwithDRR)}{\%damagecurrentsituation} \quad (8.1)$$

With a zero (0%) value meaning that the DRR measure had no benefit when compared to the current situation, while 100% indicates total risk prevention by the modelled DRR.

DRR assessment, including climate change, can be performed ex-ante, years or decades before the potential consequence, allowing for the development of the best possible DRR solution, and further testing it as time passes and new data and knowledge arises. With such an approach, the preparedness level for climate change impacts at coastal areas can be much higher than the current one, defining the most effective solution. It also allows defining at which time the solution should be implemented to optimise resources (see Fig. 8.2). Not all potential measures to minimise climate change impacts will need to be implemented at the same time or at the nearest future. Continuity of the use of models for the evaluation of DRR effectiveness would allow the definition of local timeframes for coastal management, defining the optimal management approach at each moment within the next decades.

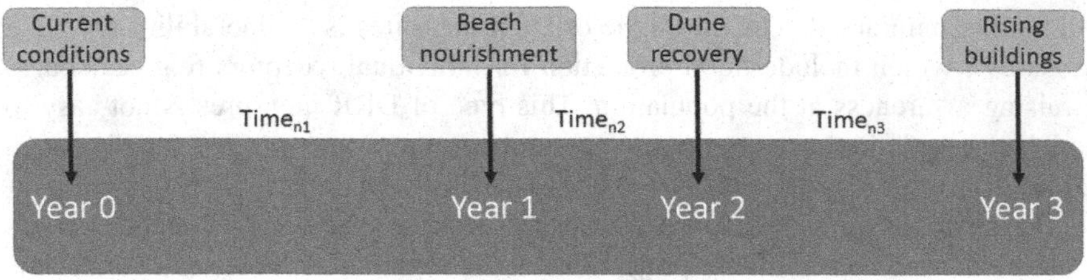

Fig. 8.2 Hypothetical timeframe for the optimal implementation of selected DRR measures, as a function of the evolution of climate change drivers

Bottlenecks and Future Developments

Modelling limitations, namely, for the more complex process-based models, include the lack of high-quality quantitative validation due to lack of data, particularly relating to water discharge, water velocities and inundation extent (Ferreira et al. 2018). Limitations on application also include the difficulty of a wide direct use by end-users, including coastal managers. The first limitation is being solved by the ongoing and increasing improvement on data access (and quality) worldwide, including online access to coastal morphology and wave/water level series, but an extra effort on field measurements is still required. To obviate the second limitation, a higher interaction will be needed on the transfer of knowledge from the coastal scientific community to end-users.

The applicability of the exposed models to define DRR measures, including climate change, will require a better determination of the vulnerability relationships between the indicators used to define damage or impact (e.g. water discharge, shoreline retreat) and the exposed elements (e.g. type of building). At the local level, to better define the vulnerability, it will be required a historical analysis of the damage promoted by other (past) events as learning lessons for the future. An extended analysis of the effectiveness of DRR measures should also be implemented, by including aspects linked to the economic value, ecosystem services and socio-cultural characteristics of each region. A cost–benefit analysis, including ecosystem services, is highly advised, to be sure that implemented DRR measures are not only effective but also less costly. Indirect impacts are also often not accounted on the use of these models. However, the full assessment of impacts should include all cascade effects. For instance, the expected increase on coastal flooding events, associated with climate change, will cause extended cascade effects, by stopping the economy of those coastal areas more often and for a longer period. Those effects are still rarely evaluated but should be incorporated in future impacts assessments.

Conclusions

Climate change and direct consequences, like sea-level rise or changes on storminess, have been predicted for the last decades, at global, regional and local levels. It is also known that these drivers will lead to a change on impacts at coastal areas, with a general trend to increasing hazards and derived consequences. High-resolution morphodynamic numerical models also exist and have been tested, validated and improved, at coastal areas, for the last decades. A high diversity of disaster risk reduction measures have also been implemented all over the world over the last decades. It is, however, not yet common to see the joint use of morphodynamic models that integrate climate change predictions, morphodynamic models and disaster risk reduction measures to define future coastal management actions, their optimisation and their potential effectiveness against climate change impacts. This work proposes a generic approach to be followed when testing adaptation and risk reduction measures for the future, considering climate change, including the application of an effective index, among other assessments. This proposal aims to contribute to the extended use of validated coastal models, encompassing a great range of modelled conditions, including climate change, enhancing disaster preparedness for effective risk reduction at coastal areas.

Acknowledgements This work was supported by the European Union 7th Framework Programme through the grant to RISC-KIT ("Resilience increasing Strategies for Coasts—Toolkit"), contract no. 603458, and the Portuguese Science and Technology Foundation through the grants attributed to project EW-COAST (ALG-LISBOA-01-145-FEDER-028657) and CIMA (UID/MAR/00350/2013).

References

Bates PD, De Roo APJ (2000) A simple raster-based model for floodplain inundation. J Hydrol 236:5477

Brown S, Nicholls RJ, Pardaens AK, Lowe JA, Tol RSJ, Vafeidis AT, Hinkel J (2019) Benefits of climate-change mitigation for reducing the impacts of sealLevel rise in G-20 countries. J Coast Res 35(4):884–895

Cumiskey L, Priest S, Valchev N, Viavattene C, Costas S, Clarke J (2018) A framework to include the (inter)dependencies of Disaster Risk Reduction measures in coastal risk assessment. Coast Eng 134:81–92

van Dongeren A, Ciavola P, Martinez G, Viavattene C, Bogaard T, Ferreira O, Higgins R, McCall R (2018) Introduction to RISC-KIT: resilience-increasing strategies for coasts. Coast Eng 134:2–9

Ferreira O, Plomaritis TA, Costas S (2017) Process-based indicators to assess storm induced coastal hazards. Earth-Sci Rev 173:159–167

Ferreira O, Plomaritis TA, Costas S (2019) Effectiveness assessment of risk reduction measures at coastal areas using a decision support system: findings from Emma storm. Sci Total Environ 657:124–135

Ferreira O, Viavattene C, Jiménez JA, Bolle A, das Neves L, Plomaritis TA, McCall R, van Dongeren AR (2018) Storm-induced risk assessment: evaluation of two tools at the regional and hotspot scale. Coast Eng 134:241–253

Hinkel J, Lincke D, Vafeidis AT, Perrette M, Nicholls RJ, Tol RSJ, Marzeion B, Fettweis X, Ionescu C, Levermann A (2014) Coastal flood damage and adaptation costs under 21st century sea-level rise. PNAS 2014(111):3292–3297

Jäger WS, Christie EK, Hanea AM, den Heijer C, Spencer T (2018) A Bayesian network approach for coastal risk analysis and decision making. Coast Eng 134:48–61

Larson M, Erikson L, Hanson H (2004) An analytical model to predict dune erosion due to wave impact. Coast Eng 51:675–696

Mendoza ET, Jiménez JA (2006) Storm-induced beach erosion potential on the Catalonian coast. J Coast Res SI 48:81–88

Plomaritis TA, Costas S, Ferreira Ó (2018) Use of a Bayesian Network for coastal hazards, impact and disaster risk reduction assessment at a coastal barrier (Ria Formosa, Portugal). Coast Eng 134:134–147

Roelvink D, Reniers A, van Dongeren AP, de Vries JVT, McCall R, Lescinski J (2009) Modelling storm impacts on beaches, dunes and barrier islands. Coast Eng 56:1133–1152

Vousdoukas MI, Mentaschi L, Voukouvalas E, Bianchi A, Dottori F, Feyen L (2018) Climatic and socioeconomic controls of future coastal flood risk in Europe. Nat Clim Chang 8:776–780

Vousdoukas MI, Mentaschi L, Voukouvalas E, Verlaan M, Jevrejeva S, Jackson LP, Feyen L (2018) Global probabilistic projections of extreme sea levels show intensification of coastal flood hazard. Nat Commun 9:2360

Vousdoukas MI, Ranasinghe R, Mentaschi L, Plomaritis TA, Athanasiou P, Luijendijk A, Feyen L (2020) Sandy coastlines under threat of erosion. Nat Clim Change 10:260–263

Vousdoukas MI, Mentaschi L, Hinkel J, Ward PJ, Mongelli I, Ciscar J, Feyen L (2020b) Economic motivation for raising coastal flood defenses in Europe. Nat Commun 11:2119

Chapter 9
A Model-Based Response Surface Approach for Evaluating Climate Change Risks and Adaptation Urgency

Timothy R. Carter and Stefan Fronzek

Abstract We present a new approach to advance methods of climate change impact and adaptation assessment within a risk framework. Specifically, our research seeks to test the feasibility of applying impact models across sectors within a standard analytical framework for representing three aspects of potential relevance for policy: (i) sensitivity—examining the sensitivity of the sectors to changing climate for readily observable indicators; (ii) urgency—estimating risks of approaching or exceeding critical thresholds of impact under alternative scenarios as a basis for determining urgency of response; and (iii) response—determining the effectiveness of potential adaptation and mitigation responses. By working with observable indicators, the approach is also amenable to long-term monitoring as well as evaluation of the success of adaptation, where this too can be simulated. The approach focuses on impacts in climate-sensitive sectors, such as water resources, forestry, agriculture or human health. It involves the construction of impact response surfaces (IRSs) based on impact model simulations, using sectoral impact models that are also capable of simulating some adaptation measures. We illustrate the types of analyses to be undertaken and their potential outputs using two examples: risks of crop yield shortfall in Finland and impact risks for water management in the Vale do Gaio reservoir, Portugal. Based on previous analyses such as these, we have identified three challenges requiring special attention in this new modelling exercise: (a) ensuring the salience and credibility of the impact modelling conducted and outputs obtained, through engagement with relevant stakeholders, (b) co-exploration of the capabilities of current impact models and the need for improved representation of adaptation and (c) co-identification of critical thresholds for key impact indicators and effective representation of uncertainties. The approach is currently being tested for five sectors in Finland.

T. R. Carter (✉) · S. Fronzek
Finnish Environment Institute (SYKE), Helsinki, Finland
e-mail: tim.carter@syke.fi

S. Fronzek
e-mail: stefan.fronzek@syke.fi

C. Kondrup et al. (eds.), *Climate Adaptation Modelling*, Springer Climate,
https://doi.org/10.1007/978-3-030-86211-4_9

Keywords Risk assessment · Probabilistic · Impact model · Shared socioeconomic pathways · Finland

Introduction

The Finnish Climate Act (no. 609/2015) specifies that the government should approve a national adaptation plan for climate change at least once every 10 years. The first adaptation plan to 2022 has the following three objectives (MMM 2014, p. 4): "(a) adaptation [should be] integrated into the planning and activities of both the various sectors and their actors, (b) the actors [should have] access to the necessary climate change assessment and management methods and (c) research and development work, communication and education and training [should enhance] the adaptive capacity of society, [develop] innovative solutions and [improve] citizens' awareness on climate change adaptation". A national risk assessment for key sectors was published in 2018 along with a governance model for organising future assessments (Hildén et al. 2018). This recommends that climate change risk assessments to support adaptation policies be integrated into existing assessments contributing to the National Security Strategy for Society. This planned enhancement of Finnish adaptation policy mirrors similar initiatives for regular risk assessments reported in other European countries (EEA 2018).

Impacts and Adaptation in a Risk Framework

In this short note, we propose a systematic method of climate change impact and adaptation assessment designed to be conducted at national or sub-national scale within a risk framework. Specifically, we seek to test the feasibility of applying impact models across sectors within a standard analytical framework for representing three aspects of potential relevance for policy: (i) *sensitivity*—examining the sensitivity of the sectors to changing climate for readily observable indicators; (ii) *urgency*—estimating risks of approaching or exceeding critical thresholds of impact under alternative scenarios to help determine urgency of response and (iii) *response*—determining the effectiveness of potential adaptation and mitigation responses. By working with observable indicators, the approach is also amenable to long-term monitoring as well as evaluation of the success of adaptation, where this too can be simulated.

Modelling Impacts and Adaptation

Numerical impact models are important research tools for evaluating climate change risks in support of decision-making. They are often deployed to quantify uncertainties in potential impacts across a range of climate projections and other scenario assumptions. However, their use in decision-making is still fairly limited, in part because of reluctance to engage stakeholders in the co-development of models in order to demonstrate a case for their applicability, salience and trustworthiness (EEA 2017).

This observation is of particular importance when simulating adaptation. A recent review found the treatment of adaptation in impact models intended for land use and water management to be fragmented and often simplistic, failing to recognise that adaptation is a complex human process framed by uncertainties and constraints (Holman et al. 2019). The review lists sixteen suggestions for future improvements in simulations of climate change adaptation (equally transferable to other sectors too). Four of these improvements are of especial relevance in the approach outlined below:

- embracing scenario uncertainty rather than seeking most likely futures for optimal solutions,
- working with stakeholders and decision-makers to better understand the triggers and goals of adaptation policies and measures,
- including adaptations that take advantage of climate change rather than simply responding to adverse impacts, and
- considering adaptation alongside mitigation within an integrated climate policy framework.

An additional aspect to highlight is the importance for model outcomes to relate to real-world evidence of adaptation (Berrang-Ford et al. 2011). This services a demand for monitoring and evaluation of adaptation by national and local policymakers and potentially for the Global Stocktake (Tompkins et al. 2018).

Objectives and Research Questions

The approach seeks to accomplish six aims. These are to:

(1) Work with sectoral experts and stakeholders to co-select demonstration indicators that are observable and for which impacts and adaptation measures can be simulated using models,

(2) Construct impact response surfaces (IRS—see section below) to depict the modelled sensitivity of indicators to climate and socioeconomic drivers across a plausible range of perturbations in different geographical regions, accounting for uncertainty where applicable,

(3) Estimate the evolving likelihood of exceeding stakeholder-defined thresholds of impact and hence the urgency to act under alternative scenarios during the twenty-first century,
(4) Simulate historical climate change impacts to compare to observed impacts and adaptation,
(5) Examine the modelled effectiveness of adaptation and mitigation for ameliorating adverse impacts or exploiting beneficial impacts, and
(6) Develop protocols for model analysis and for effective visualisation and dissemination of results to feed into national risk assessments.

To address these specific aims, a modelling methodology is being designed to ensure some commonality of approach between sectoral applications. This requires collective decisions to be agreed at an early stage of an assessment through a process of co-design across sectors and between researchers and stakeholders. Protocols can then be agreed that modellers can use to carry out the model simulations needed for addressing the major research questions, which include (in relation to the above objectives):

(1) What impact indicators that are of relevance to stakeholders can be readily quantified and modelled using the IRS approach?
(2) How sensitive are these indicators to changes in climate and other key driving variables?
(3) By when and with what likelihood will critical thresholds of impact be exceeded in the future under alternative scenarios of socioeconomic and climate change?
(4) Is there evidence that these thresholds have already been exceeded in the past?
(5) How effective is adaptation and mitigation at reducing risks of exceeding critical thresholds?
(6) Can a common approach to analysis be operationalised for use in national risk assessments?

Approach

Operationalising the IPCC Risk Framework

The approach builds on the premise that the IPCC Risk Framework can be operationalised by modelling impacts of climate change for key indicators co-selected with stakeholders across a range of sectors, relating these impacts to impact thresholds and simulating the effectiveness of adaptation and mitigation in ameliorating key risks. The IPCC risk framework depicts climate change risk, R, as

$$R = PI = f(H, E, V) \tag{9.1}$$

where H is the hazard, describing aspects of the climate that may induce adverse impacts (i.e. changes in the mean and/or variability, including extreme events). E

is exposure, which is the proximity of humans, ecosystems, infrastructures or other economic, social or cultural assets that could be adversely affected by the hazard. *V* is vulnerability, defined as the predisposition of the exposed elements to be adversely affected. The risk term, *R*, in Eq. (9.1) is sometimes interpreted in terms of potential impact, *PI* (IPCC 2014). Hence, risk (potential impact) is a function of the hazard posed by climate change (using climate model projections, for example, based on representative concentration pathways—RCPs), which can be moderated through mitigation, and of vulnerabilities (exposure, susceptibility and coping capacity) that are mediated by future socioeconomic trends (based, for example, on shared socioeconomic pathways—SSPs) and can be adjusted through adaptation.

Impact Models

The approach focuses on impacts in climate-sensitive sectors (e.g. water resources, forestry, agriculture, human health and winter recreation) and is built around simulations with a set of impact models. These models should be capable of simulating relevant indicators for the selected sectors and of incorporating options for adapting to climate change. They may operate at a variety of scales (e.g. site, catchment, regional, national) but can potentially be scaled up to the geographical units of relevance in the assessment.

Impact Response Surfaces

The method being applied for examining changing risk and the urgency for adaptation across different sectors involves the construction of impact response surfaces (IRSs) based on impact model simulations (Jones 2000). IRSs depict the response of an impact variable to changes in two explanatory variables as a plotted surface (see Fig. 9.1a). They can be used to evaluate responses to any scenario of the drivers that falls within the sensitivity range of the plot, hence providing a systematic, "scenario-neutral" analysis of impacts (Prudhomme et al. 2010) that does not rely on the arbitrary and opportunistic use of scenario simulations. While the approach may lack the internal consistency between variables that can be represented in detailed scenarios, our experience suggests that there are few cases where such simplification may produce radically different responses from those found for detailed scenarios, though such differences can of course be tested. IRSs also provide an opportunity to test model performance across a wide range of conditions, including those that may lie outside the conventional application of many models.

The IRS method has been increasingly applied during the past decade for illustrating impact model sensitivity to climate variables (e.g. temperature and precipitation) in sectors such as agriculture, hydrology and ecosystems (e.g. Poff et al. 2016; Fronzek et al. 2019). Pertinent to this study, IRSs have been combined with

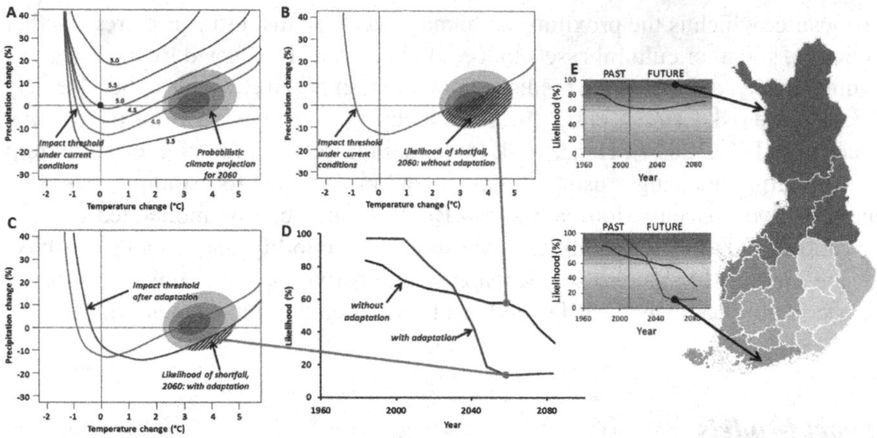

Fig. 9.1 Features of impact response surfaces (IRSs) and their potential application in estimating climate change risks (see text for explanation)

probabilistic representations of future climate (e.g. Räisänen and Ruokolainen 2006) enabling estimates of the likelihood of certain pre-specified impact thresholds being crossed (Pirttioja et al. 2019). They have also been used to model responses to adaptation measures (e.g. Ruiz-Ramos et al. 2018).

Illustrative Results: Risk Assessment

Combining the IRS method with probabilistic projections of driving variables to estimate future climate-related risks is a novel approach with, as yet, limited uptake. Most applications have been in water resource management or agriculture, and we illustrate two such cases below using recent examples from Finland and Portugal. However, its practical merits have yet to be demonstrated for other sectors and are the subject of ongoing research.

Risks of Crop Yield Shortfall in Finland

In this example, results from a published site-based modelling study (Pirttioja et al. 2019) are schematised and extended to illustrate (hypothetically) how an IRS analysis can be used to estimate regional risk (Fig. 9.1). In A, yield sensitivity to temperature and precipitation perturbations relative to a reference climate (black dot) is shown as contours, with a threshold yield level indicated in red. Such a threshold yield could be determined with stakeholders. A probabilistic representation of projected climate at some time in the future is superimposed on the IRS (darker shades indicate higher probability). The likelihood that the future climate would cause a yield shortfall can be estimated as the area on the climate surface where yields on the IRS lie below the threshold (B). Using similar estimates for several time periods into the future, a graph can be plotted showing the changing likelihood of crop failure (black line in D). The effectiveness of an adaptation measure under perturbed climate (e.g. changing to a different crop cultivar) can also be explored, by repeating the IRS analysis for the simulated adaptation (C) and constructing a new likelihood curve (blue line in D).

Impact Risks for Water Management in the Vale Do Gaio Reservoir, Portugal

The Vale do Gaio reservoir in the dry region of southern Portugal is used for irrigation of rice cultivation in the area. We constructed IRSs of the water inflow to the reservoir, the irrigation water demand for the current rice cultivation and the Water Exploitation Index (WEI; ratio between irrigation demand and runoff) with catchment-scale hydrological and irrigation models, and as adaptation options also for four other crops (winter wheat, olive trees, sunflower, corn) with a smaller water demand (Fronzek et al., in prep.). IRSs were then combined with probabilistic climate change projections similar to the crop yield example shown in Fig. 9.1. Results showed large risks of an inflow decrease throughout the twenty-first century and an increased risk of water scarce conditions from *extremely unlikely* (<5% probability) in the period 2011–2040 to *virtually certain* (>99% probability) for RCP8.5 by 2071–2100 under the current rice cultivation. Switching to crops with a smaller water demand, on the other hand, provided a potential even to increase the area under irrigation, but with an enhanced sensitivity to changes in rainfall compared to the current rice cultivation.

Regional Risks and the Urgency for Action

With appropriate data for model input, calibration and testing, the site-based approach to risk assessment can potentially be extended to national scale. We illustrate this for the same crop yield example (Fig. 9.1). A national analysis of the risk of yield shortfall might involve construction of equivalent likelihood curves for representative sites in different regions, and mapped for a given time period with the level of risk colour coded (E). This method of risk mapping shares characteristics with the reasons for concern used in the IPCC assessments (e.g. IPCC 2014) or traffic light warning systems for defining levels of risk and could be a useful device for indicating the level of urgency for action, whether by adaptation to ameliorate the risk or mitigation to avert the hazard. Note that estimates of likelihood can also be applied to climatic conditions already experienced historically, potentially allowing for comparison of risk estimates with actual observation of impacts being monitored in different regions.

The approach has an added advantage for regular risk assessments that it can be updated as new scenarios appear, without needing to re-run the underlying impact models.

Conclusions and Recommendations

Based on previous analyses such as those illustrated above, we conclude that there are three challenges requiring special attention in this new model-based approach to risk assessment: (a) ensuring the salience and credibility of the impact modelling conducted and outputs obtained, through engagement with relevant stakeholders, (b) co-exploration of the capabilities of current impact models and the need for improved representation of adaptation and (c) co-identification of critical thresholds for key impact indicators and effective representation of uncertainties. The approach is currently being tested in five sectors at national scale in Finland (https://www.syke.fi/projects/adapt-first).

Acknowledgements The authors acknowledge funding from the Adapt-FIRST project of the Academy of Finland (decision 330915) and the European Commission FP7 project IMPRESSIONS (Grant Agreement No. 603416).

References

Berrang-Ford L, Ford JD, Paterson J (2011) Are we adapting to climate change? Global Environ Change 21:25–33

EEA (2017) Climate change adaptation and disaster risk reduction in Europe: enhancing coherence of the knowledge base, policies and practices. 15/2017, Europ Environ Agency 172 pp

EEA (2018) National climate change vulnerability and risk assessments in Europe, 2018. 1/2018, Europ Environ Agency 79 pp

Fronzek S, Carter TR, Pirttioja N, et al (2019) Determining sectoral and regional sensitivity to climate and socio-economic change in Europe using impact response surfaces. Reg Environ Change 19:679–693

Fronzek S, Pirttioja N, Honda Y, et al (in prep) Estimating impact likelihoods from probabilistic projections of climate and socio-economic change using impact response surfaces (in preparation)

Hildén M, Haavisto R, Harjanne A, et al (2018) Climate resilient Finland—a governance model for organizing weather and climate risk assessments (in Finnish). Prime Minister´s Office, Publication 44/2018:67 pp

Holman IP, Brown C, Carter TR, Harrison PA, Rounsevell M (2019) Improving the representation of adaptation in climate change impact models. Reg Environ Change 19:711–721

IPCC (2014) Climate change 2014: synthesis report. Contribution of Working Groups I, II and III to the Fifth Assessment Report of the Intergovernmental Panel on Climate Change. In: Pachauri RK, Meyer LA (eds) Core writing team. Geneva, Switzerland, 151 pp

Jones RN (2000) Analysing the risk of climate change using an irrigation demand model. Climate Res 14:89–100

MMM (2014) Finland's national climate change adaptation plan 2022. Publications of the Ministry of Agriculture and Forestry 5b/2014, 40 pp

Pirttioja N, Palosuo T, Fronzek S, Räisänen J, Rötter RP, Carter TR (2019) Using impact response surfaces to analyse the likelihood of impacts on crop yield under probabilistic climate change. Agr Forest Meteorol 264:213–224

Poff NL, Brown CM, Grantham TE, et al (2016) Sustainable water management under future uncertainty with ecoengineering decision scaling. Nat Clim Change 6:25–34

Prudhomme C, Wilby RL, Crooks S, Kay AL, Reynard NS (2010) Scenario-neutral approach to climate change impact studies: application to flood risk. J Hydrol 390:198–209

Räisänen J, Ruokolainen L (2006) Probabilistic forecasts of near-term climate change based on a resampling ensemble technique. Tellus 58A:461–472

Ruiz-Ramos M, Ferrise R, Rodríguez A, et al (2018) Adaptation response surfaces for managing wheat under perturbed climate and CO_2 in a Mediterranean environment. Agric Syst 159:260–274

Tompkins EL, Vincent K, Nicholls RJ, Suckall N (2018) Documenting the state of adaptation for the global stocktake of the Paris agreement. Wires Clim Change. https://doi.org/10.1002/wcc.545

Chapter 10
Use of Vegetation for Landslide Risk Mitigation

Bjørn Kalsnes and Vittoria Capobianco

Abstract Landslide risk management involves several activities, modelling being a required premise for most of them. Modelling of climate-induced landslides include both the analysis of the triggering process, i.e. static slope stability analysis and dynamic propagation (run-out) analysis. These analyses are vital for mapping purposes, as well as for selection of effective means to reduce the landslide risk when this exceeds a certain value of tolerance. With the prospect of increasing rainfall duration and intensity in parts of Europe, the need for further development of modelling tools is evident. In recent years, the use of Nature-Based Solutions (NBS) for mitigation of natural hazards has further demonstrated the need for developing the modelling tools. The use of vegetation as NBS is increasingly being used for erosion protection and shallow landslide mitigation. For slope stability analyses, the use of vegetation makes the modelling more complex for a number of reasons, mostly linked to the influence of vegetation on both the soil–atmosphere interaction (i.e. rainfall interception, evapotranspiration) and the soil hydro-mechanical properties. All effects that are difficult to model due to lack of knowledge and to large variations in time and space. Even though there is an increasing activity in the geotechnical environment to incorporate the effects of vegetation in the modelling for quantifying the change in slope stability (i.e. calculate slope safety factor), the status is far from being at the level of traditional landslide modelling tools. More efforts are therefore needed in the years to come to demonstrate that the use of vegetation as a viable and effective measure in landslide risk mitigation management can be verified in a more quantifiable manner.

Keywords Landslides · Nature-based solutions · Mitigation · Vegetation · Slope stability modelling

B. Kalsnes (✉) · V. Capobianco
NGI, Oslo, Norway
e-mail: bgk@ngi.no

V. Capobianco
e-mail: vic@ngi.no

C. Kondrup et al. (eds.), *Climate Adaptation Modelling*, Springer Climate,
https://doi.org/10.1007/978-3-030-86211-4_10

Introduction

Landslide risk management in the context of climate change has been a profiled study for more than a decade. Many studies have shown that a change in rainfall duration and intensity will cause an increase in natural water-induced phenomena, such as floods, soil erosion and landslides in large parts of Europe, with damaging effects on people, infrastructure, housings and the environment. The need for a proper landslide risk management strategy is therefore significant at all scales, namely, national, regional and local. A premise for sound landslide risk management is modelling of triggering and run-out phenomena, to determine location and extent of potential landslides and thus the selection of appropriate risk reduction measures.

In recent years, there has been an increasing focus on the use of Nature-Based Solutions (NBS), both with regard to urban and rural development, and for disaster risk reduction. This paper presents the challenges related to the use of NBS for landslide mitigation purposes. The question is simply: how can we verify that the use of NBS is an effective measure for mitigating a landslide problem for a detailed case, and simultaneously being not harmful to the environment? The focus of the paper will be on modelling of slope stability with the use of vegetation. What are the effects of vegetation in reducing the probability of landslide occurrence, and how do we model these effects?

Climate-Induced Landslides

Landslides Risk in View of Climatic Changes: Relevant Past and On-Going Projects

The effects of climate change on the landslide risk have been a major concern for many years. The need to protect people and property with a changing pattern of landslide hazard and risk caused by climate change and changes in demography was the main motivation for the FP7 research project 'SafeLand' (2009–2012) on landslide risk in Europe (Nadim and Kalsnes 2014). In the SafeLand project, considerable effort was done on developing models for the prediction of precipitation-induced landslides. One of the conclusions was that the thresholds for landslide triggering are affected by long-term precipitations in areas that are covered by deep deposits of fine-grained soils, while they are controlled by short-term precipitations in areas with shallower deposits with coarse-grained soils. For shallow landslides, the soil–atmosphere interaction is a major factor influencing the slope stability. Various geotechnical stability programmes are able to model these effects as slope top boundary conditions, taking into account the pore pressure development and general soil behaviour characteristics.

The main aim of any management strategy is to reduce the landslide risk to acceptable levels when found necessary. This can be done using structural and/or non-structural measures (for instance, early warning systems). Structural means may include measures to hinder the landslide to develop, thus stopping the triggering phase, or measures to reduce the run-out effects of a landslide already taking place. As a follow-up of an activity in SafeLand, the Norwegian Research Center Klima 2050 has developed a web-based tool LaRiMiT (Landslide Risk Mitigation Toolbox, https://www.larimit.com) aimed at assisting decision-makers to select an appropriate mitigation measure for a given landslide problem (Uzielli et al. 2017). More than 80 various measures are identified in LaRiMiT, most of them relevant for rainfall-induced landslides. Out of a total of 11 categories of landslide mitigation measures, 2 categories and a total of approximately 15 measures imply NBS measures or hybrid measures (combination of NBS and traditional 'grey' measures). Most of the measures are relevant for erosion control and shallow landslides.

Slope Stability Modelling

The landslide modelling normally implies two phases, one is the geotechnical static slope stability analysis and the other is the dynamic propagation analysis (run-out). The first serves for the hazard analysis and the latter serves for both hazard analysis and the identification of hazard scenarios, as input for estimating the consequences of a certain landslide event. The use of vegetation is not yet sufficiently addressed in neither of them. In this paper, the focus is on the geotechnical modelling of static slope stability, i.e. hazard analysis.

Two main modelling principles are used for geotechnical static slope stability analyses: (i) the limit equilibrium methods (LEM) and (ii) the finite element methods (FEM). The principal difference between these two methods is that LEM is based on static equilibrium, while the FEM uses the stress–strain relationships or the constitutive law, to simulate the mechanical behaviour of the soil. The LEM method identifies potential failure mechanisms and derives factors of safety. Among the various LEM methods available, those most used satisfy both force and moment equilibriums. FEM requires the definition and the use of complex constitutive models for all materials, especially for describing the soil behaviour. Different constitutive laws may be used, for example, linear elastic–perfectly plastic, linear elastic-hardening plastic laws. In both cases, the modelling of the soil behaviour is the key to reliable results, thus detailed field and laboratory tests are required for defining input parameters.

Nature-Based Solutions (NBS)

Nature-Based Solutions for Climate-Related Challenges: European Strategy

Nature-Based Solutions (NBS) is a collective term for solutions that are based on natural processes and ecosystems to solve different types of societal challenges. Of particular interest is mitigation and adaptation strategies to address climate-related challenges. The use of NBS has several advantages beyond their primary goals, such as preventing natural hazards. IUCN (2017) points out the breadth of benefits the use of NBS can include: (a) increasing biodiversity; (b) long-term stability; (c) ecological management both 'upstream and downstream'; (d) direct societal benefits; (e) local governance.

A first milestone in the establishment of NBS was the World Bank's report *Biodiversity, Climate Change and Adaptation: Nature-Based Solutions from the World Bank Portfolio* (World Bank 2008). In recent years, NBS has received increased attention, not least as a result of the European Commission (EC) investing considerable resources in building up European competitive advantage in this field. The EC has, indeed, established a clear strategy of Europe being a main actor in the development and use of NBS for various climate-related societal challenges. A large number of research programmes have been launched since 2014; one of them is related to use of NBS for hydrometeorological risk reduction (EC 2017). These studies incorporate the use of NBS for landslide risk mitigation, which also includes the need for proper modelling tools. However, the latter are far from being at the level of traditional landslide modelling tools, even though the interest is increasing internationally. More efforts are needed in the years to come to be able to handle in a quantitative manner the use of NBS for landslide mitigation.

Climate change will cause a change of rainfall patterns and intensity in large parts of Europe. This will lead to an increased probability for rainfall-induced landslides with high destructive potential for exposed infrastructure. In order to reduce the societal risk associated with climate change and enhanced precipitation, NBS can represent a sustainable, efficient and cost-effective approach. NBS have been increasingly applied to design new resilient landscapes and cities with beneficial outcomes for the environment, the society and human well-being.

Use of NBS in Landslide Risk Mitigation

In the recent years, a large variation of NBS measures were proposed for mitigating natural hazards. Some of them are grounded in the Ecosystem-Disaster Risk Reduction (Eco-DRR) with the aim to achieve sustainable and resilient development (Estrella and Saalismaa 2013). Sutherland et al. (2014) identified almost 300 NBS-specific measures for natural hazards mitigation and for agricultural problems. For landslide and erosion protection, most of these measures involved the use of vegetation. Arce-Mojica et al. (2019) made a similar study, focussing on the NBS measures for reducing the risk of shallow landslides. They performed a systematic literature review to ascertain the extent to which vegetation is identified as a controlling factor and the targeting of NBS for landslide risk reduction. They concluded that despite there has been an important increase in the number of articles dealing with NBS approaches for shallow landslides mitigation; science appears to be lagging behind compared to the promotion of NBS in international and policy arenas. There is a need for further research, both related to a most suitable selection of vegetation species in different forest ecosystems and biogeographical regions, which is essential for a successful mitigation, and to the potential negative effects of vegetation as a shallow landslide triggering factor.

Modelling of Slope Stability Using Vegetation

Effects of Vegetation on Landslide Protection

Several studies have identified both positive and negative effects of using vegetation for landslide protection (Stokes et al. 2014; NVE/NGI 2015; Krzeminska et al. 2019). The major findings are that the use of vegetation for landslide protection have two positive effects and one potential negative effect: (i) the strength of the soil increases due to roots and binding of soil layers, (ii) the pore water pressure is reduced due to plant's uptake and canopy cover, (iii) vegetation may destabilize slopes in connection with strong winds (this is valid only for trees). These are all effects that may be modelled, but as the studies show there are a lot of uncertainties related to this aspect. Examples of challenges with regard to modelling include the following:

- The undrained shear strength depends on the type of roots, the position of the main roots network and the season of the year.
- The effect of reduced soil water content and induced soil suction is highly uncertain and can vary considerably from case to case, also in relation to the distribution and vegetation density along the slope.

Expected Development Within Landslide Modelling Using Vegetation

Landslide modelling when including the vegetation contribution in slope stability analyses will be more complex, due to the coupled effect that they provide to the soil: (i) hydrological, through the soil–vegetation–atmosphere interaction and (ii) mechanical, through the root–soil interaction.

Figure 10.1 shows a methodological approach which takes into account the vegetation contribution in the slope stability modelling. The approach consists of two main parts: (i) hydrological modelling, to assess the pore water pressure regime and (ii) slope stability modelling, to assess the safety factor. As input data, hydrometeorological analysis implies the collection of current meteorological data (e.g. rainfall intensity, wind, temperature, relative humidity), or the analysis of potential future climate scenarios, to be used to feed the hydrological model. It is important to stress that precipitation events are often linked to the triggering of landslides, but it is the change in pore water pressures that leads a slope to fail (Toll et al. 2011). As it concerns, the input data related to the soil, many soil parameters as well as hydraulic processes (water fluxes) are function of the vegetation. A tentative to categorize the effects of vegetation on the input data has been done on the base of whether they are function of the root features (mostly density, architecture and depth) or the canopy (type of aboveground vegetation).

For the hydrological modelling and the evaluation of the pore water pressure regime in the ground, the hourly rainfall is an essential input to the water flux, while both the roots and the aboveground vegetation features influence the processes and the soil parameters. Some challenges related to the definition of these relationships are as follows:

Soil hydraulic properties: The hydraulic conductivity of the soil strongly depends on the type of roots (coarse or fine) and their age (i.e. young roots or decaying roots). Some preliminary functions were proposed to model the effect of roots on the change of soil hydraulic conductivity, but they have been included so far only in analytical analyses (Ni et al. 2018). However, recent studies have found that

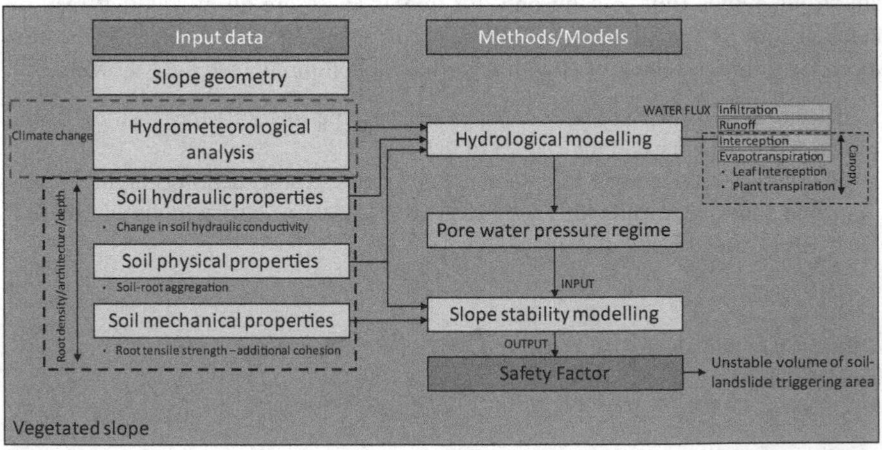

Fig. 10.1 Methodological approach for slope stability modelling including vegetation—soil parameters and processes influenced by the presence of vegetation (root density/architecture/depth and canopy area)

hydraulic conductivity of the soil can change also with time as the roots develop and grow (Capobianco et al. 2020).

Soil physical properties: As roots occupy the pores, they tend to change also the soil void ratio. How much they change the soil unit weight needs additional studies.

Interception: This accounts the rainwater intercepted by the vegetation that does not infiltrate into the soil. Such factor is strongly affected by the canopy area and the parameters such as the Leaf Area Index (LAI).

Evapotranspiration: Most of the hydrological models calculate the potential evapotranspiration with the equation proposed by Penman-Monteith (Allen et al. 1998), in which the potential transpiration given by the vegetation is usually function of the LAI and the soil cover fraction.

Once the pore water pressure regime is assessed, this is used as input for the slope stability modelling, where only the root features are considered to influence the mechanical properties of the soil. The effect of roots on the soil mechanical properties has been extensively studied and understood from the perspective of geotechnical engineering: the root tensile strength provides additional cohesion to the soil with a magnitude depending on the tensile strength and the root density. It is evident that the soil–vegetation–atmosphere interaction is complex and requires both the knowledge of the root features and how the vegetation is developed aboveground.

Challenges Related to Legislation

Vegetation has traditionally been used for erosion protection in many areas of Europe. The positive effects of use of vegetation for shallow landslides have also been widely recognized. However, due to challenges with regard to quantifying these positive effects, use of vegetation is often overseen as a practical measure for landslide protection. When it comes to building and construction, Eurocode standards need to be followed in many European countries. This implies that a minimum safety factor of 1.4 (ratio between stabilizing forces and driving forces) needs to be demonstrated for local slope stability analyses. In such cases, the need for better modelling is needed as the tools available at present is not sufficient for demonstrating properly in quantitative manner the positive effects of use of vegetation for landslide protection.

Conclusions and Recommendations

Use of vegetation as a viable and effective measure in landslide risk mitigation management needs to be documented in a more quantifiable manner. The effect of vegetation is complex and varies with time, type of soil and atmospheric conditions. A methodological approach to include the vegetation in slope stability modelling in hazard prone areas is herein proposed, where the hydrological and mechanical reinforcement provided by the vegetation on the soil properties are classified whether they are root-related or canopy-related. Some key challenges in this respect are as follows:

Modelling the vegetation effects on slope stability needs many parameters related to the vegetation features which are strongly time-dependent. Moreover, vegetation features differ from species to species. There is a need to understand how to consider the vegetation growth effects.

Only one safety factor is calculated as average. Small-scale effects of vegetation on slope stability are not calculated. However, vegetation may not be distributed homogeneously, thus there is a need to consider time-spatial variation of vegetation effects on a slope (Stokes et al. 2014).

Climate change may alter the precipitation scheme dramatically in many areas, with more intense rainfall combined with more dry periods. The need for combined efforts in local instrumentation and modelling development is pronounced.

This study on landslide modelling is focussed on addressing the effects of vegetation on rainfall-induced landslides. However, climate change may also increase the frequency of droughts, which lead to tree mortality and forest fires. Possible extreme events like these, which still imply the vegetation, need to be studied with regard to the initiation of extreme surface runoff and flash floods due to heavy rainfall.

References

Allen RG, Pereira LS, Raes D, Smith M (1998) Crop evapotranspiration-Guidelines for computing crop water requirements-FAO Irrigation and drainage paper 56. Fao, Rome 300(9):D05109

Arce-Mojica T, Nehren U, Sudmeier-Rieux K, Miranda P, Anhuf D (2019) Nature-based solutions (NbS) for reducing the risk of shallow landslides; where do we stand? Int J Disaster Risk Reduct 41 (2019). https://doi.org/10.1016/j.ijdrr.2019.101293

Capobianco V, Cascini L, Cuomo S, Foresta V (2020) Wetting-drying response of an unsaturated pyroclastic soil vegetated with long-root grass. Environ Geotech 1–18

EC (2017) Large-scale demonstrators on nature-based solutions for hydro-meteorological risk reduction. SC5-08

Estrella M, Saalismaa N (2013) Ecosystem-based disaster risk reduction (Eco-DRR): an overview. The role of ecosystems in disaster risk reduction. United Nations University Press, Tokyo, 332

International union for conservation of nature, IUCN, 2017. The IUCN Global Programme 2013–16, adopted by the IUCN World Conservation Congress, September 2012

Krzeminska D, Kerkhof T, Skaalsveen K, Stolte J (2019) Effect of riparian vegetation on stream bank stability in small agricultural catchments. CATENA 172:87–96

Nadim F, Kalsnes B (2014) Progress of living with landslide risk in Europe. In: Sassa K et al (ed) Keynote paper to 3rd World Landslide Forum, Beijing, June 2014. Landslide science for a Safer Geoenvironment, vol 1. Springer International Publishing, Switzerland. https://doi.org/10.1007/9783-319-04999-1-1

Ni JJ, Leung AK, Ng CWW, Shao W (2018) Modelling hydro-mechanical reinforcements of plants to slope stability. Comput Geotech 95:99–109

NVE/NGI (2015) Oppsummeringsrapport for skog og skredprosjektet. Samanstilling av rapportar frå prosjektet. Rapport 92/2015 (In Norwegian)

Sutherland WJ, Gardner T, Bogich TL, Bradbury RB, Clothier B, Jonsson M, Kapos…and Dicks, L.V. V (2014) Solution scanning as a key policy tool: identifying management interventions to help maintain and enhance regulating ecosystem services. Ecol Soc 19(2):3. https://doi.org/10.5751/ES-06082-190203

Stokes A, Douglas GB, Fourcaud T, Giadrossich F, Gillies C, Hubble T, Kim JH, Loades, Walker LR, et al (2014) Ecological mitigation of hillslope instability: ten key issues facing researchers and practitioners. Plant Soil 377:1–23. https://doi.org/10.1007/s11104-014-2044-6

Toll DG, Lourenco SDN, Mendes J, Gallipoli D, Evans FD, Augarde CE, Mancuso C (2011) February. Geological society of London, Soil suction monitoring for landslides and slopes

Uzielli M, Kalsnes B, Choi JC (2017) A web-based landslide risk mitigation portal. In: Proceedings, 4th World Landslide Forum, Ljubljana, Slovenia, June 2017

World Bank (2008) Biodiversity, climate change and adaptation: nature-based solutions from the world bank portfolio. Washington, DC

Chapter 11
Modelling to Evaluate Climate Resilience of Crop Rotations Under Climate Change

Kurt Christian Kersebaum

Abstract Diversification of crop rotations is considered as an option to increase the resilience of European crop production under climate change. Although crop rotation design and management has been identified as an important measure to adapt to and mitigate climate change, most studies on climate change impact or adaptation so far use single-year simulations and/or single crop assessments. Crop response to various management options within a growing season is generally taken into account by most crop models. However, if simulations neglect processes and fluxes between growing seasons and potential carry-over effects related to agronomic management, the long-term sustainability of adaptation and mitigation strategies cannot be properly evaluated. Therefore, the integrated assessment of impacts, adaptation and mitigation options under current and future climatic conditions requires a continuous long-term analysis of crop sequences to take into account carry-over effects as in real conditions. The present paper provides information on crop rotation aspects, which should be considered in modelling, presents the current state of modelling for climate impact assessment, address points of uncertainty and missing aspects in modelling and draws an outlook on potential future developments with special emphasis on crop rotations. In conclusion, crop models require suitable experimental data to parameterize additional crops, which were so far not sufficiently investigated to cope with multiple opportunities in crop rotations.

Keywords Agro-ecosystem model · Crop rotation · Climate resilience · Carry-over effects · Model gaps

Introduction

Agriculture is among the most vulnerable sectors affected by climate change. Additionally, it has strong interactions to other sectors, e.g. water resource management and mitigation of greenhouse gas emissions. Observed climate trends showed an

---

K. C. Kersebaum (✉)
Leibniz Centre for Agricultural Landscape Research, Müncheberg, Germany
e-mail: ckersebaum@zalf.de

C. Kondrup et al. (eds.), *Climate Adaptation Modelling*, Springer Climate,
https://doi.org/10.1007/978-3-030-86211-4_11

increase of land temperature by ~1.5 °C compared to pre-industrial level, an increase of the frequency, intensity and duration of heat waves, while precipitation showed a spatially differentiated picture across Europe. Warming has prolonged the thermal growing season and frost-free period for crops across all of Europe since 1980s. Extended growing seasons facilitate the introduction of new crops or the expansion to higher latitudes and altitudes. However, the probability of multiple adverse climate events during a growing season including heavy rains and storms has increased and cropping systems have already been affected by climate change in terms of higher yield variability and crop loss risk but also in terms of adaptation, e.g. new varieties, diversification of crops or shifted cultivation dates. Projected climate trends are likely to enhance the risk posed by extreme weather events under climate change scenarios. To cope with increasing global food demand adaptation of agricultural production systems is required to minimize the risk of yield losses and exploit new chances from changing climatic conditions, while meeting other sustainability goals such as environmental impacts and efficient resource use. Diversification of crop rotations is considered as one option to increase the resilience of European crop production under climate change. Therefore, the design and management of crop rotations play an important role for the adaptation of cropping systems.

Main Effects of Crop Rotations and Their Management

The sequence of crops in crop rotations plays a significant role in resource use efficiency. Inefficient use of nitrogen in fertilizers and manures enhance N surplus and losses in many cropping systems across Europe, leading to nitrate pollution of groundwater and surface waters and to enhanced greenhouse gas emissions, especially N_2O. Leaky periods in crop rotations can be prevented by the use of intermediate crops (e.g. catch crops), which retain N in the rooting zone and carry it over to the subsequent crop. Implementation of legumes as main or intermediate crops provides additional nitrogen to the following crop and needs to be considered in operational fertilization decisions. Additionally, winter cover crops prevent soils from erosion. However, the establishment of intermediate crops depends on water availability at sowing and their water consumption may reduce water availability for the following crop. Their effect on pest and diseases might be divers and it has to be considered if winter cover crops might be a host for diseases, suppress weeds or act as enemy crops, e.g. oil radish against nematodes. Crop sequence effects on yield can persist for 3–4 years in dry years or semi-arid environments as a result of water and nutrient legacies. Such legacy effects also includes inoculum survival and subsequent infestations of crops with fungal diseases. On long term, crop and soil management systems are known to change the storage of soil organic matter. Other mechanisms associated with crop sequences, e.g. effects on soil structure and soil physical processes and their interactions with roots are still not well understood.

Current State of Model-Based Climate Change Impact Assessment

Agro-ecosystem models considering the complex interactions in the atmosphere-plant–soil system are essential tools to assess the impact of projected climate change on various ecosystem services, including crop production, environmental effects such as water use and pollution, GHG emissions, resource use efficiency and long-term effects on soil properties, e.g. soil carbon stocks and to evaluate potential adaptation options under these multiple aspects. Their main advantage is the opportunity to conduct high number of simulations under various site conditions within a relatively short time.

During the last decade large international research consortia like the European JPI FACE knowledge hub MACSUR or the global AgMIP activity performed modelling studies to assess climate change impacts for main staple crops using large model ensembles. Their main message was, that no single model performed best across all site conditions and that the mean or median of a model ensemble mostly provided the best estimate (Asseng et al. 2013). Moreover, the ensemble of crop models showed a high variability indicating the uncertainty related to the model or its user, respectively. The majority of agricultural climate change impact assessment studies were focussed on few specific variables, mainly crop yields and considered only the main climatic factors such as temperature (Asseng et al. 2015) or drought effects either solely or in combination (Webber et al. 2018a). However, this approach has limitations, considering that crop growth and yield are affected by the interactive impacts of multiple climate change factors and biophysical processes. Webber et al. (2018b) suggested to replace air temperature by simulated canopy temperature to consider the interaction between heat and drought stress. However, the interaction between heat and drought seems to be genotype dependent and is still not sufficiently understood (Rötter et al. 2018).

Although the effect of elevated CO_2 is considered in most of the models, their response on photosynthesis and water use varied substantially (Asseng et al. 2013). Kersebaum and Nendel (2014) analysed the effect of using three different CO_2 response functions in combination with a dynamic CO_2 response of stomatal resistance within the model HERMES across 21 regions in Germany on wheat yield, groundwater recharge and nitrogen leaching under current and projected climate. Model results for wheat yields differed by 5.5–11.6% among the three methods. Moreover, results showed a strong dependency on site conditions (soil and groundwater level) regarding the vulnerability against climate change. Diverse results regarding the beneficial effect of transpiration reduction through elevated CO_2 emphasized, that the statement of higher CO_2 stimulation, when crops are under water stress, cannot be applied universally. While algorithms seem to be applicable for different C3 crops, models showed a weak performance regarding the CO_2 effect on C4 crops under reduced water availability in a FACE experiment (Durand et al. 2018).

Response to extreme weather events other than heat and drought, e.g. heavy rain, storm, hail, frost or water logging are less considered in most crop models (Rötter et al. 2018). Although yield reductions due to lodging were reported to be 31–80% in wheat, 4–65% in barley, 37–40% in oats, 5–20% in maize and 5–84% in rice, lodging is usually not considered in commonly used crop models. Moreover, efforts on the effects of extremes were focussed mainly on the three main staple crops wheat, maize and rice, while studies on other crops are rare (Rötter et al. 2018).

Although crop rotation design and management has been identified as an important measure to adapt to and mitigate climate change, most studies on climate change impact or adaptation so far use single-year simulations and/or single crop assessments (Webber et al. 2018a). However, if simulations neglect to include year-to-year changes in initial soil conditions of water and nutrient availability related to agronomic management, adaptation and mitigation strategies cannot be properly evaluated. Therefore, the integrated assessment of impacts, adaptation and mitigation options under current and future climatic conditions requires continuous long-term simulations of crop sequences (e.g. Kollas et al. 2015) to take into account carry-over effects as in real conditions. Although Kollas et al. (2015) used more than 300 years of experimental data for their model inter-comparison on crop rotations, the performance of models regarding crop yields improved only slightly when continuous simulation was used instead of annual resetting to standardized initial conditions. Lack of pronounced carry-over effects was because the effect of nutrient transfer on yields of the following crop was often masked by a high fertilization level and most of the sites were located in humid environments, where soil water mostly reached field capacity during winter. However, assessing the effect on other target variables, e.g. nitrogen balance compounds is only possible when using a continuous simulation since most emissions occur during the fallow periods (Yin et al. 2020).

Kollas et al. (2015) stated that models showed a weak performance mainly on crops where only few data were available for a proper calibration and modellers were less experienced. While main crops are usually well parameterized, the data base for not widely used crops or non-commercial crops such as potatoes, sugar beets or catch crops was often not sufficient for a solid parameterization. This underlines the request of Rötter et al. (2018) to extend research to crops other than the main staple crops.

Although some studies have conducted simulations for crop rotations mainly investigating effects of catch crops on nitrate leaching or N_2O emissions (Yin et al. 2020; Gillette et al. 2018) under current conditions, only few studies looked at crop rotations under climate change scenarios (Hlavinka et al. 2014).

Conclusions and Recommendations for Model Improvement

Modelling of crop rotations require models that cope with a large variety of crops and management operations. As stated above, capability of models to simulate complex crop rotations is limited by the availability of suitable field datasets to parameterize and calibrate various crops such as oilseeds, pulses or beet crops, and those, which are of less economic value, but may contribute to environmental benefits such as catch crops. While data for model validation might be available from standard field trials, data requirements for model calibration are demanding since they require a high data density and proper description of the boundary and site conditions (Kersebaum et al. 2015). Therefore, they are rarely available for crops beside the main staple crops. Also suitable data on crop failure following extreme events are rare since experiments are usually cancelled after such events. Crop responses to several abiotic stress factors are still not fully understood and multifactorial manipulative experiments are thus crucial for a proper model-based assessment of plants growth and development under current and expected future environmental conditions (Rötter et al. 2018). This requires to evaluate model performance across different output variables beyond crop yield, e.g. soil water and nitrogen status and eventually severity of disease or pest damages (Kersebaum et al. 2015).

While some models are capable to simulate crop rotation effects in terms of carry-over effects of water and nitrogen, other effects like the exploration of rooting depth by previous crops for the following crop are rarely considered (e.g. Seidel et al. 2019). Many models are also lacking on suitable approaches to consider mixed cropping systems or under-sown catch crops and the competition among crops, but also with weeds.

Tillage and mulching effects are mainly considered using empirical relations, if considered at all, and only a few models are using process-based approaches (Yang et al. 2019). While no-till or minimum tillage is still propagated as a measure to sequester soil organic carbon, worldwide meta-analyses of tillage experiments cannot universally confirm this statement. Moreover, the effect of tillage seems to depend on site conditions, e.g. combined soil–climate impact. Therefore, the implementation of process-based approaches is required to reflect the site-specific short- (e.g. soil water balance) and long-term (soil organic carbon) responses on different tillage and residue management practices. This becomes even more important before the background of glyphosate use discussion, which is the usual alternative for tillage.

Crop losses by pest and diseases are rarely considered in crop models. Within MACSUR and AgMIP first attempts have been made to implement generic damage mechanisms to assess crop loss by pest and diseases, which builds on early modelling efforts during the 1980s (Bregaglio et al. 2021). However, following the roadmap of Donatelli et al. (2017) models are under development, which consider the interaction between crops and pests and diseases through the link between crop models and pest and disease models to cope with future changes of biotic pressures under climate change. This may also include the consideration of crop rotation effects on initial infection probability.

Acknowledgements The author acknowledge support by the Federal Ministry of Education and Research (BMBF), Germany (031B0039C and 031B0513I) through MACSUR2 and I4S and the project "SustES—Adaptation strategies for sustainable ecosystem services and food security under adverse environmental conditions" (CZ.02.1.01/0.0/0.0/16_019/0000797) by the Ministry of Education, Youth and Sports of Czech Republic.

References

Asseng S, Ewert F, Rosenzweig C, Jones JW, Hatfield JL, Ruane A, Boote KJ, Thorburn P, Rötter RP, Cammarano D, Brisson N, Basso B, Martre P, Aggarwal PK, Angulo C, Bertuzzi P, Biernath C, Challinor A, Doltra J, Gayler S, Goldberg R, Grant R, Heng L, Hooker J, Hunt T, Ingwersen J, Izaurralde C, Kersebaum KC, Müller C, Naresh Kumar S, Nendel C, O'Leary G, Olesen JE, Osborne TM, Palosuo T, Priesack E, Ripoche D, Semenov M, Shcherbak I, Steduto P, Stöckle C, Stratonovitch P, Streck T, Supit I, Tao F, Travasso M, Waha K, Wallach D, White J, Williams JR, Wolf J (2013) Quantifying uncertainties in simulating wheat yields under climate change. Nat Clim Change 3:827–832

Asseng S, Ewert F, Martre P, Rötter RP, Lobell DB, Cammarano D, Kimball BA, Ottman MJ, Wall GW, White JW, Reynolds MP, Alderman PD, Prasad PVV, Aggarwal PK, Anothai J, Basso B, Biernath C, Challinor AJ, De Sanctis G, Doltra J, Fereres E, Garcia-Vila M, Gayler S, Hoogenboom G, Hunt LA, Izaurralde RC, Jabloun M, Jones CD, Kersebaum KC, Koehler A-K, Müller C, Naresh Kumar S, Nendel C, O'Leary G, Olesen JE, Palosuo T, Priesack E, Eyshi Rezaei E, Ruane AC, Semenov MA, Shcherbak I, Stöckle C, Stratonovitch P, Streck T, Supit I, Tao F, Thorburn P, Waha K, Wang E, Wallach D, Wolf J, Zhao Z, Zhu Y (2015) Rising temperatures reduce global wheat production. Nat Clim Change 5(2):143–147

Bregaglio S, Willocquet L, Kersebaum KC, Ferrise R, Stella T, Ferreira TB, Pavan W, Asseng S, Savary S (2021) Comparing process-based wheat growth models in their simulation of yield losses caused by plant diseases. Field Crops Res 265:108108

Donatelli M, Magarey RD, Bregaglio S, Willocquet L, Whish JPM, Savary S (2017) Modelling the impacts of pests and diseases on agricultural systems. Agric Syst 155:213–224

Durand J-L, Delusca K, Boote K, Lizaso J, Manderscheid R, Weigel H-J, Ruane AC, Rosenzweig C, Jones JW, Ahuja L, Anapalli S, Basso B, Baron C, Bertuzzi P, Biernath C, Deryng D, Ewert F, Gaiser T, Gayler S, Heinlein F, Kersebaum KC, Kim S-H, Müller C, Nendel C, Olioso A, Priesack E, Ramirez Villegas J, Ripoche D, Rötter RP, Seidel SI, Srivastava A, Tao F, Timlin D, Twine T, Wang E, Webber H, Zhao Z (2018) How accurately do maize crop models simulate the interactions of atmospheric CO_2 concentration levels with limited water supply on water use and yield? Eur J Agron 100:67–75

Gillette K, Malone RW, Kaspar TC, Ma L, Parkin TB, Jaynes DB, Fang QX, Hatfield JL, Feyereisen GW, Kersebaum KC (2018) N loss to drain flow and N_2O emissions from a corn-soybean rotation with winter rye. Sci Total Environ 618:982–997

Hlavinka P, Trnka M, Kersebaum KC, Cermák P, Pohanková E, Orság M, Pokorný E, Fischer M, Brtnický M, Žalud Z (2014) Modelling of yields and soil nitrogen dynamics for crop rotations by HERMES under different climate and soil conditions in the Czech Republic. J Agric Sci 152:188–204

Kersebaum KC, Nendel C (2014) Site-specific impacts of climate change on wheat production across regions of Germany using different CO_2 response functions. Eur J Agron 52:22–32

Kersebaum KC, Boote KJ, Jorgenson JS, Nendel C, Bindi M, Frühauf C, Gaiser T, Hoogenboom G, Kollas C, Olesen JE, Rötter RP, Ruget F, Thorburn PJ, Trnka M, Wegehenkel M (2015) Analysis and classification of data sets for calibration and validation of agro-ecosystem models. Environ Model Softw 72:402–417

Kollas C, Kersebaum KC, Nendel C, Manevski K, Müller C, Palosuo T, Armas-Herrera CM, Beaudoin N, Bindi M, Charfeddine M, Conradt T, Constantin J, Eitzinger J, Ewert F, Ferrise R, Gaiser T, Garcia de Cortazar-Atauri I, Giglio L, Hlavinka P, Hoffmann H, Hoffmann MP, Launay M, Manderscheid R, Mary B, Mirschel W, Moriondo M, Olesen JE, Öztürk I, Pacholski A, Ripoche-Wachter D, Roggero PP, Roncossek S, Rötter RP, Ruget F, Sharif B, Trnka M, Ventrella D, Waha K, Wegehenkel M, Weigel H-J, Wu L (2015) Crop rotation modelling – a European model intercomparison. Eur J Agron 70:98–111

Rötter RP, Appiah M, Fichtler E, Kersebaum KC, Trnka M, Hoffmann MP (2018) Linking modelling and experimentation to better capture crop impacts of agroclimatic extremes—a review. Field Crop Res 221:152–156

Seidel S Gaiser T, Kautz T, Bauke SL, Amelung W, Barfus K, Ewert F, Athmann M (2019) Estimation of the impact of precrops and climate variability on soil depth differentiated spring wheat growth and water, nitrogen and phosphorus uptake. Soil Till Res 195:104427

Webber H, Ewert F, Olesen JE, Müller C, Fronzek S, Ruane A, Bourgault M, Martre P, Ababaei B, Bindi M, Ferrise R, Finger R, Fodor N, Gabaldón-Leal C, Gaiser T, Jabloun M, Kersebaum KC, Lizaso JI, Lorite I, Manceau L, Moriondo M, Nendel C, Rodríguez A, Ruiz Ramos M, Semenov MA, Siebert S, Stella T, Stratonovitch P, Trombi G, Wallach D (2018) Diverging importance of drought stress for maize and winter wheat in Europe. Nat Commun 9:4249

Webber H, White JW, Kimball BA, Ewert F, Asseng S, Rezaei EE, Pinter PJ Jr, Hatfield JL, Reynolds MP, Ababaei B, Bindi M, Doltra J, Ferrise R, Kage H, Kassie BT, Kersebaum KC, Luig A, Olesen JE, Semenov MA, Stratonovitch P, Ratjen AM, LaMorte RL, Leavitt SW, Hunsaker DJ, Wall GW, Martre P (2018) Physical robustness of canopy temperature models for crop heat stress simulation across environments and production conditions. Field Crop Res 216:75–88

Yang W, Feng G, Adeli A, Kersebaum KC, Jenkins JN, Li PF (2019) Long-term effect of cover crop on rainwater balance components and use efficiency in the no-tilled and rainfed corn and soybean rotation system. Agric Water Manage 219:27–39

Yin X, Kersebaum, KC, Beaudoin N, Constantin J, Chen F, Louarn G, Manevski K, Hoffmann M, Kollas C, Armas-Herrera C, Baby S, Bindi M, Dibari C, Ferchaud F, Ferrise R, Garcia de Cortazar-Atauri I, Launay M, Mary B, Moriondo M, Özturk I, Ruget F, Sharif B, Wachter-Ripoche D, Olesen JE (2020) Uncertainties in simulating N uptake, net N mineralization, soil mineral N and N leaching in European crop rotations using process-based models. Field Crops Res 255:107863

Chapter 12
Dynamic Flood Risk Modelling in Human–Flood Systems

Heidi Kreibich and Nivedita Sairam

Abstract Effective flood risk management is highly relevant for advancing climate change adaptation. It needs to be based on risk modelling that considers the dynamics, complex interactions and feedbacks in human–flood systems. In this regard, we review recent advancements in understanding, quantifying and modelling changes in risk and its drivers. A challenge for integrating human behaviour in dynamic risk assessments and modelling is the combined consideration of qualitative and quantitative data. Advancements in this respect are (1) the compilation and analysis of comprehensive qualitative and quantitative data on flood risk changes in case studies following the paired event concept; (2) the integration of qualitative and quantitative data into socio-hydrological models using Bayesian inference; and (3) the coupling of hydrological flood risk models with behaviour models in socio-hydrological modelling systems. We recommend to further develop these approaches and use more such process-based, dynamic modelling also for large-scale flood risk analyses. These approaches are increasingly feasible due to significant improvements in computational power and data science.

Keywords Socio-hydrology · Stylized models · System-of-systems models · Agent-based models

Introduction

Flood risk assessment and management are highly relevant for advancing climate change adaptation, since floods cause very large amounts of material damage and casualties worldwide (Kundzewicz et al. 2014). From all natural hazards, they affected the largest number of people (>2 billion) globally in the period 1998–2017

H. Kreibich (✉) · N. Sairam
GFZ German Research Centre for Geosciences, Section Hydrology, Telegrafenberg, Potsdam, Germany
e-mail: heidi.kreibich@gfz-potsdam.de

N. Sairam
e-mail: nivedita.sairam@gfz-potsdam.de

C. Kondrup et al. (eds.), *Climate Adaptation Modelling*, Springer Climate,
https://doi.org/10.1007/978-3-030-86211-4_12

and they caused €52 billion in overall losses in Europe in the period 1998–2009. Due to climate change and increasing exposure, e.g. via urbanization, the risk of flooding is expected to increase in the future (Kundzewicz et al. 2014). However, data and modelling driven studies also show that effective adaptation including flood risk management has a high potential to counteract the effect of climate change (Kreibich et al. 2017; Metin et al. 2018).

There is general agreement that flood risk, as well as its components (hazard, exposure and vulnerability) are dynamic, and should be treated as such (Vorogushyn et al. 2018). Hazard is defined as the potential occurrence of an event that may cause adverse effects on social elements, while exposure is defined as the presence of people, livelihoods, environmental services and resources, infrastructure or economic, social or cultural assets in places that could be adversely affected by physical events. Vulnerability is defined generically as the propensity or predisposition to be adversely affected. Finally, impacts, e.g. direct damage such as fatalities or economic damage, represents risk. Adaptation aims to reduce the overall risk, which can be done by reducing the hazard (i.e. the frequency/magnitude of flooding, e.g. via structural protection measures such as retention basins or levees), the exposure of people and properties or their vulnerability to flooding. However, changing one of those risk components may lead to unexpected behaviour of the system as a whole, resulting in phenomena like the levee effect, i.e. the increase of exposure and vulnerability behind levees due to the non-occurrence of flooding (Di Baldassarre et al. 2018).

Due to continuous feedbacks across the human–flood system, risk-based decision-making requires understanding, quantifying and projecting changes in risk under an integrated, systems framework (Barendrecht et al. 2020; Vorogushyn et al. 2018). In this regard, this review is focused on approaches to quantify the temporal dynamics of risk and its drivers as well as to project flood risk changes in the future. It draws particularly from two preceding studies: the opinion paper 'How to improve attribution of changes in drought and flood impacts' (Kreibich et al. 2019) and the review paper 'A dynamic framework for flood risk' (Barendrecht et al. 2017). Thus, state-of-the-art empirical data-driven knowledge and modelling methods are discussed.

Empirical Data-Driven Knowledge

Aggregated flood damage data at the event level, available from several global, regional and national databases, are used for trend analyses (Bouwer 2011). However, due to the event level and large spatial scales of analyses, the studies cannot provide insights into processes (Bouwer 2011). Empirical flood risk data on the micro-scale in case studies, available from participatory studies, surveys, official statistics or open access data sources are valuable for gaining process understanding in respect to the dynamics of risk (e.g. Sairam et al. 2019). However, long-term analyses in local case studies are only rarely possible. Thus, the recently developed paired event concept is an important advancement (Kreibich et al. 2017, 2019). It consists of analysing changes in risk and its components (hazard, exposure, vulnerability) as well as processes and interactions based on paired events in the same catchment or region, irrespective of the time between events (Kreibich et al. 2017). It is not limited to one pair of events, but the more events that are considered for the same region, the better. The paired event concept is analogous to the concept of 'paired-catchment studies', which is a well-established concept in hydrology (Kreibich et al. 2019).

Temporal trend analyses on flood damage data detect a clear increase in damage (Bouwer 2011). Most of these studies find that the observed increase is due to societal change and economic development. An effect on damage from changes in flood hazard due to climate change has hardly been observed until now. However, exposure and vulnerability are largely influenced by human interventions such as flood protection and their interaction and influence on risk can only be roughly accounted for over time (Bouwer 2011). Thus, it is hypothesized that an increase in flood hazard is counteracted by a decrease in vulnerability, e.g. via effective flood risk management, including protection, early warning and preparation (Jongman et al. 2015). A decrease in vulnerability seems to have occurred at the global level since about 1980, which is reflected in decreasing mortality and losses as a share of population and gross domestic product exposed to river flooding (Jongman et al. 2015). On national scale, it is reported, e.g. that in Bangladesh, vulnerability towards flooding has decreased strongly since 1974, which seems to be due to substantial improvements in flood risk management (Mechler and Bouwer 2015). An empirical analysis of eight paired event case studies around the world showed that an observed reduction in flood impacts was driven mainly by reductions in vulnerability (Fig. 12.1). Such detailed case study-based analyses revealed that vulnerability can be positively influenced by integrated flood risk management, which complements structural protection with non-structural solutions, e.g. private precaution, land-use planning and insurance (Kreibich et al. 2017). However, vulnerability can also be negatively influenced by changes in building materials, increasing dependence on critical infrastructure, or changes in business processes. For instance, recent reports by insurers emphasize that floods cause tremendous losses, particularly to modern buildings with good thermal insulation and innovative building materials. While these buildings perfectly fulfil the requirements of energy-saving standards that are important to mitigate climate change in the long run, it seems that such constructions tend to increase average building losses due to their high susceptibility to flooding (Kreibich et al. 2019). In summary, knowledge about changes in vulnerability and risk, particularly their drivers and driver interactions is scarce so that more monitoring and empirical data analyses are necessary, including new data sources such as satellite data and social media.

Modelling Changes in Flood Risk

Since data-driven knowledge is limited to inferences derived from past trends, modelling approaches are necessary for projecting future trends or developing future scenarios for flood risk. Therefore, modelling approaches that consider drivers of risk along with their interactions and feedbacks are a necessary step forward for

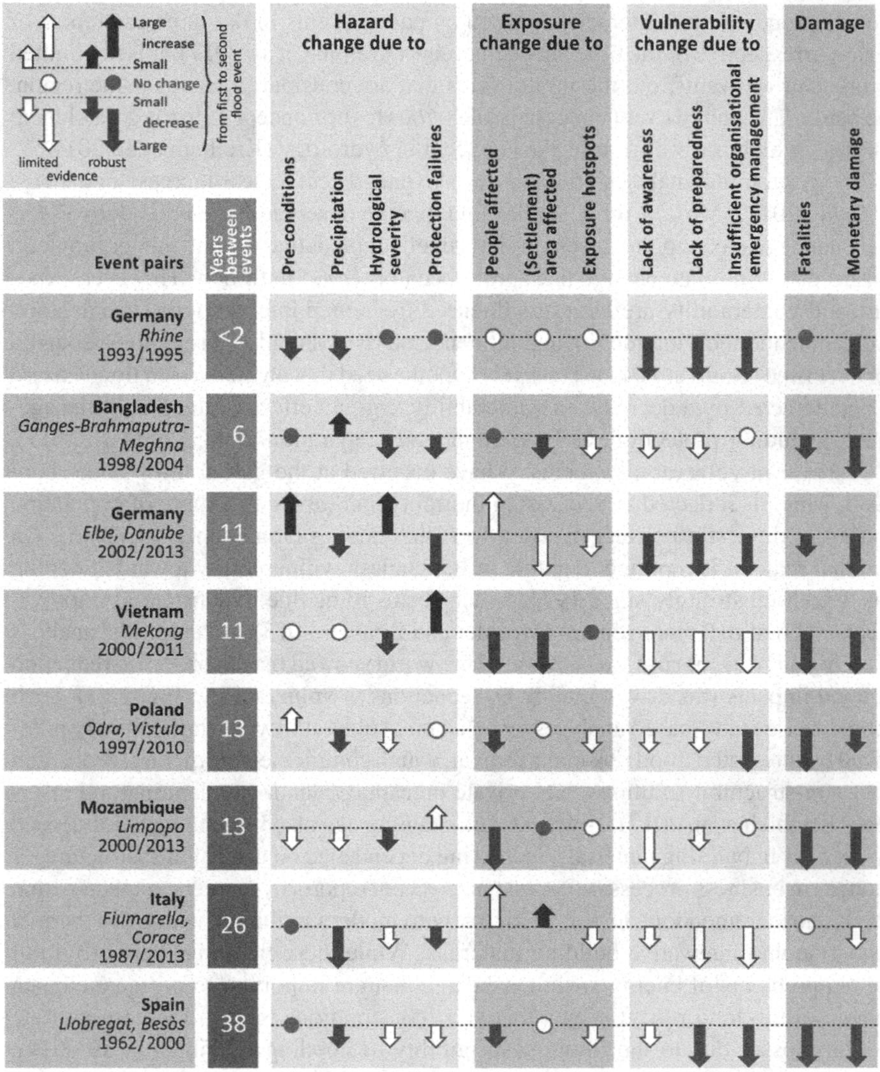

Fig. 12.1 Analysis of eight paired flood events showing the difference in primary drivers of change in flood risk and fatalities and economic losses between the first flood event, which serves as the baseline, and the second event (published by Kreibich et al. 2017)

adaptation. In this section, three categories of state-of-the-art socio-hydrological modelling approaches that aim to project trends in flood risk are discussed: stylized models (SYMs), system-of-systems models (SSMs) and agent-based models (ABMs) (Barendrecht et al. 2017). The implementation of these approaches is strongly influenced by the spatial scale of risk assessment, availability of expert knowledge and empirical data.

Modelling Approaches

SYMs capture system characteristics based on a set of processes, which are simplified into a set of differential equations (Viglione et al. 2014). These models are examples of a top-down approach, and are relatively straightforward to implement. They can be used to interpret the general characteristics of the system. For example, a local flood risk SYM is implemented in the city of Dresden, Germany (Barendrecht et al. 2019). This is a lumped model with simplified representations of relationship between flood experience, awareness, preparedness and damage processes. The use of Bayesian inferencing in the SYM allows hydrologists and social scientists to introduce their degree of belief in certain processes as priors. Further, this opens up the possibility to integrate empirical qualitative and quantitative data from recorded events as evidence (Barendrecht et al. 2019). In this case, data for the case study of Dresden, over a period of 200 years, were used to estimate the model parameters through Bayesian inference. As such, the inferences from the model are helpful to understand the general nature of human–flood feedbacks prevalent in the specific case study. Thus, changes in flood risk based on general system characteristics can be quantitatively estimated using process-based SYMs.

The SSMs are developed by coupling relevant detailed individual models that capture different processes within the system. These models are spatially explicit and as such capable of producing risk scenarios that are most relevant to the regional/local scale. SSMs often include assumptions regarding some components of the system and possible synthetic scenarios, as well. An example of an SSM relevant to flood risk is the regional flood model (RFM) implemented in the Elbe catchment in Germany (Falter et al. 2015; Metin et al. 2018). The advancements in the hydrology and hydraulics along with the increase in computational capabilities have enabled continuous simulation of the flood risk chain (Falter et al. 2015). This is a significant advancement in comparison to the previous simple assembly of local, static inundation maps (Metin et al. 2020). However, in comparison with the hazard component, there is a lot of scope for advancing quantification of changes in vulnerability, considering human–flood dynamics. Though this study is based on an SSM which consists of coupled well-researched hydrological, hydraulic and multivariable damage models (Fig. 12.2), synthetic adaptation scenarios define the feedbacks within the human–flood system (Falter et al. 2015).

ABMs capture the characteristics of individual components in the system (agents), their interactions and feedbacks. The overall system characteristics may be inferred based on this. This is an example of a bottom-up approach. The ABMs may be process-oriented where the behaviour of each agent is modelled based on behavioural theories such as protection motivation theory, expected utility theory or prospect theory (Haer et al. 2017). Additionally, evidence from empirical data may be used to determine the behaviour or update the model using Bayesian inferencing (Haer et al. 2017).

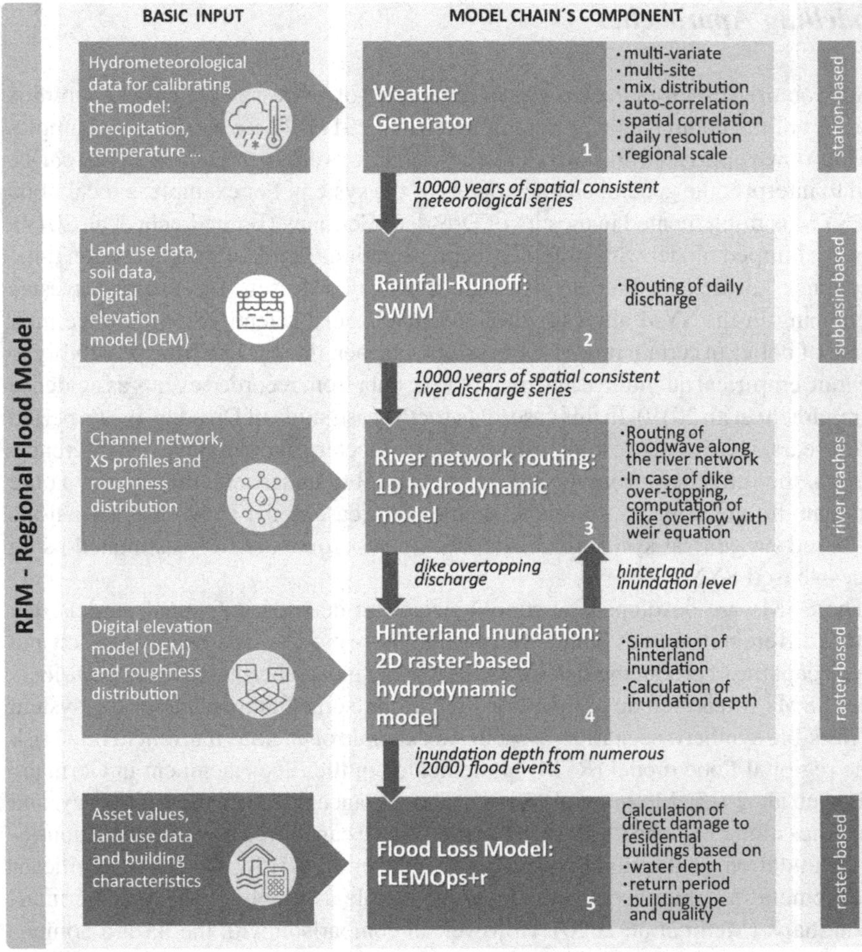

Fig. 12.2 Components and data requirements of the Regional Flood Model—RFM (adapted from Falter et al. 2015)

Role of Spatial Scale

Flood risk is stochastic and exhibits spatio-temporal variability with respect to damage processes (Sairam et al. 2019). It is possible for local/regional modelling studies to use vulnerability scenarios relevant to the case study. However, in the case of large-scale flood risk assessments at the continental or global scales, shared socioeconomic pathway scenarios are commonly used as vulnerability scenarios. As for instance for the implementation of an integrated ABM into a large-scale flood risk assessment model for the European Union (Haer et al. 2019). Recent studies derive global vulnerability projections based on comparing flood damage (losses and fatalities) against coarse indicators such as population and gross domestic product of regions (Jongman et al. 2015). More effort is needed for translating micro-level human–flood interactions and feedbacks into large-scale modelling frameworks for improved decision-making.

Conclusion and Recommendation

The dynamic modelling of flood risk considering human–flood interactions is highly relevant for an effective flood risk management and as such for advancing climate change adaptation. Important recent advancements in quantifying human–flood dynamics and integrating this knowledge into dynamic flood risk modelling are the following:

- Compiling and analysing qualitative and quantitative data about temporal changes in hazard, exposure, vulnerability and impacts and how these components are influenced by risk management in case studies advances our process understanding of human–flood systems. A promising approach is the paired event concept presented by Kreibich et al. (2017).
- Integrating qualitative and quantitative data into socio-hydrological models supports the simulation of real, long-term processes in human–flood systems. Available empirical data from recorded events can be used to calibrate and validate the models. A promising approach is using Bayesian inferences in the modelling to integrate qualitative and quantitative data as shown by Barendrecht et al. (2019).
- Coupling hydrological flood risk models with behaviour models in a socio-hydrological modelling system captures the feedback processes across the human–flood systems. A promising approach is the coupling of an ABM into a flood risk modelling system, like presented by Haer et al. (2019).

Despite these advancements, current approaches for large-scale flood risk assessments still largely ignore basic interactions and feedbacks of the human–flood systems and use too coarse data and models. Thus, we recommend to develop and use more process-based modelling systems based on more detailed and comprehensive data also for large-scale flood risk analyses, which becomes more and more feasible due to significant improvements in computational power and data science.

References

Barendrecht MH, Viglione A, Blöschl G (2017) A dynamic framework for flood risk. Water Secur 1:3–11

Barendrecht MH, Viglione A, Kreibich H, Merz B, Vorogushyn S, Blöschl G (2019) The value of empirical data for estimating the parameters of a socio-hydrological flood risk model. Water Resour Res 55(2):1312–1336

Barendrecht MH, Sairam N, Cumiskey L, Metin AD, Holz F, Priest S, Kreibich H (2020) Needed: a systems approach to improve flood risk mitigation through private precautionary measures. Water Secur 11:100080. https://doi.org/10.1016/j.wasec.2020.100080

Bouwer LM (2011) Have disaster losses increased due to anthropogenic climate change? Bull Am Meteor Soc 92(1):39–46

Di Baldassarre G, Kreibich H, Vorogushyn S, Aerts J, Arnbjerg-Nielsen K, Barendrecht M, Bates P, Borga M, Botzen W, Bubeck P, De Marchi B, Llasat C, Mazzoleni M, Molinari D, Mondino E, Mård J, Petrucci O, Scolobig A, Viglione A, Ward PJ (2018) Hess opinions: an interdisciplinary research agenda to explore the unintended consequences of structural flood protection. Hydrol Earth Syst Sci 22:5629–5637. https://doi.org/10.5194/hess-22-5629-2018

Falter D, Schröter K, Nguyen D, Vorogushyn S, Kreibich H, Hundecha Y, Apel H, Merz B (2015) Spatially coherent flood risk assessment based on long-term continuous simulation with a coupled model chain. J Hydrol 524: –193. https://doi.org/10.1016/j.jhydrol.2015.02.021

Haer T, Botzen WW, de Moel H, Aerts JC (2017) Integrating household risk mitigation behavior in flood risk analysis: an agent-based model approach. Risk Anal 37(10):1977–1992

Haer T, Botzen WW, Aerts JC (2019) Advancing disaster policies by integrating dynamic adaptive behaviour in risk assessments using an agent-based modelling approach. Environ Res Lett 14(4):044022

Jongman B, Winsemius HC, Aerts JC, De Perez EC, Van Aalst MK, Kron W, Ward PJ (2015) Declining vulnerability to river floods and the global benefits of adaptation. Proc Natl Acad Sci 112(18):E2271–E2280

Kreibich H, Di Baldassarre G, Vorogushyn S, Aerts JCJH, Apel H, Aronica GT, Arnbjerg-Nielsen K, Bouwer LM, Bubeck P, Caloiero T, Do TC, Cortès M, Gain AK, Giampá V, Kuhlicke C, Kundzewicz ZW, Llasat MC, Mård J, Matczak P, Mazzoleni M, Molinari D, Nguyen D, Petrucci O, Schröter K, Slager K, Thieken AH, Ward PJ, Merz B (2017) Adaptation to flood risk - results of international paired flood event studies. Earth's Fut 5(10):953–965. https://doi.org/10.1002/2017EF000606

Kreibich H, Blauhut V, Aerts JCJH, Bouwer LM, Van Lanen HAJ, Mejia A, Mens M, Van Loon AF (2019) How to improve attribution of changes in drought and flood impacts. Hydrol Sci J—J des Sci Hydrol 64(1):1–18. https://doi.org/10.1080/02626667.2018.1558367

Kundzewicz ZW, Kanae S, Seneviratne SI et al (2014) Flood risk and climate change: global and regional perspectives. Hydrol Sci J 59(1):1–28. https://doi.org/10.1080/02626667.2013.857411

Mechler R, Bouwer LM (2015) Understanding trends and projections of disaster losses and climate change: is vulnerability the missing link? Clim Change 133:23. https://doi.org/10.1007/s10584-014-1141-0

Metin AD, Dung NV, Schröter K, Guse B, Apel H, Kreibich H, Vorogushyn S, Merz B (2018) How do changes along the risk chain affect flood risk? Nat Hazards Earth Syst Sci 18:3089–3108. https://doi.org/10.5194/nhess-18-3089-2018

Metin AD, Dung NV, Schröter K, Vorogushyn S, Guse B, Kreibich H, Merz B (2020) The role of spatial dependence for large-scale flood risk estimation . Nat Hazards Earth Syst Sci 20:967–979. https://doi.org/10.5194/nhess-20-967-2020

Sairam N, Schröter K, Rözer V, Merz B, Kreibich H (2019) Hierarchical Bayesian approach for modelling spatio-temporal variability in flood damage processes. Water Resour Res 55(10):8223–8237

Viglione A, Di Baldassarre G, Brandimarte L, Kuil L, Carr G, Salinas JL et al (2014) Insights from socio-hydrology modelling on dealing with flood risk. J Hydrol 518:71–82. https://doi.org/10.1016/j.jhydrol.2014.01.018

Vorogushyn S, Bates PD, de Bruijn K, Castellarin A, Kreibich H, Priest S, Schröter K, Bagli S, Blöschl G, Domeneghetti A, Gouldby B, Klijn F, Lammersen R, Neal JC, Ridder N, Terink W, Viavattene C, Viglione A, Zanardo S, Merz B (2018) Evolutionary leap in large-scale flood risk assessment needed. Wiley Interdisc Rev Water 5(2):e1266. https://doi.org/10.1002/wat2.1266

Chapter 13
Climate-Fit.City: Urban Climate Data and Services

Filip Lefebre, Koen De Ridder, Katerina Jupova, Judith Köberl, Dirk Lauwaet, Antonella Passani, Jan Remund, Patrick Willems, and Katrien Witpas

Abstract The Climate-fit.City service (https://www.climate-fit.city) provides the best available scientific urban climate data and information for public and private end users operating in cities. Within the Climate-fit.City H2020 project, the benefits of urban climate information for end user communities was demonstrated, considering services in diverse domains (Climate and Health, Building Energy, Emergency Planning, Urban Planning, Active Mobility, Tourism and Cultural Heritage) to improve decision-making and to help end users to better address the consequences of climate

---

F. Lefebre (✉) · K. De Ridder · D. Lauwaet
Vlaamse Instelling voor Technologisch Onderzoek (VITO), Mol, Belgium
e-mail: filip.lefebre@vito.be

K. De Ridder
e-mail: koen.deridder@vito.be

D. Lauwaet
e-mail: dirk.lauwaet@vito.be

K. Jupova
GISAT, Prague, Czech Republic
e-mail: katerina.jupova@gisat.cz

J. Köberl
Joanneum Research, Graz, Austria
e-mail: judith.koeberl@joanneum.at

A. Passani
T6 Ecosystems S.R.L. (T6), Rome, Italy
e-mail: a.passani@t-6.it

J. Remund
Meteotest, Bern, Switzerland
e-mail: jan.remund@meteotest.ch

P. Willems
KU Leuven, Leuven, Belgium
e-mail: patrick.willems@kuleuven.be

K. Witpas
ArctiK, Brussels, Belgium
e-mail: katrien.witpas@artik.eu

C. Kondrup et al. (eds.), *Climate Adaptation Modelling*, Springer Climate,
https://doi.org/10.1007/978-3-030-86211-4_13

change at the local scale. The socio-economic impact assessment performed in the Climate-fit.City project has demonstrated that, in all the cases, there are actual and potential added values in terms of public service effectiveness, economic impacts, policy innovation and social impacts. Further impact was also revealed in terms of raising awareness by end users, policymakers and the general public about climate change. These diversified impacts offer a variegated landscape of sub-areas and stakeholders that are touched upon by each climate service.

Keywords Urban climate change · Urban heat stress · Urban flooding · Urban adaptation planning

Introduction

Urban areas are very vulnerable to climate change impacts, because of the high concentration of people, infrastructure and economic activity but also because cities tend to exacerbate climate extremes such as heat waves and flash floods. In addition, the ongoing urban expansion and the ageing of the urban population makes them particularly vulnerable. European cities are home to about 75% of the population, projected to grow to 80% in 2050. The objective of the Climate-fit.City service was to establish a service that translates the best available scientific urban climate data into relevant information for public and private end users operating in cities.

Urban areas shape their own climate, amplifying climate extremes such as excessive heat and flooding. Lauwaet et al. (2015) demonstrated that, because of the urban heat island effect, cities experience twice as many heat wave days than their rural surroundings. Moreover, towards the end of the century, the number of urban heat wave days is expected to increase by a factor of 10, from approximately 3 to 30 days per year at the end of the century under IPCC scenario RCP8.5. With respect to water, the abundance of impermeable surfaces in cities leads to inundations that are often far more intense than those occurring in rural areas (Willems et al. 2012), damaging property and infrastructure and causing economic losses arising from disrupted transportation networks.

In view of the ongoing and projected climate change, urban areas need to set up adaption processes to become less sensitive to the negative impacts of climate change. This transformation needs to be cross-sectorial as climate impacts many urban activities that are linked to each other. In this paper, we will start by describing how Climate-fit.City supports urban adaptation and its data processing methodology (Sect. "Climate-Fit.City Data") followed by a brief presentation of the Climate-fit.City service components (Sect. "Climate-Fit.City Services") and a brief reflexion on the use of the Climate-fit.City data and services and the impacts that it generates (Sect. "Conclusions").

Climate-Fit.City Data

The Climate-fit.City partners want to help cities and urbanized regions manage current and future climate impacts. The diverse consortium of climate experts, thematic sectorial experts, socio-economic experts and a professional communication partner work together with city officials to gather and integrate climate data to get a clear view of specific local challenges and co-design solutions to them. The Climate-fit.City team helps cities through the process, supports stakeholder engagement and has expertise in socio-economic impact assessments and policy design including communicating of climate impacts to citizens. The Climate-fit.City modular approach offers expertise in the following urban sectors: active mobility, building energy, emergency planning, heat and health, leisure and tourism and urban planning (Fig. 13.1). We will use results created inside the H2020 Climate-fit.City project (2017–2020) to demonstrate service components.

Users of the Climate-fit.City service are on the one hand urban administrations/institutions and on the other hand territorial entities as well as private companies supporting cities by providing customized information and services to:

- define the climate change risk by mapping and quantifying diverse impacts which increases visibility, awareness and ownership of the adaptation challenge at both the policy and citizen levels;
- support the development and adoption of adaptation strategies and actions plans;

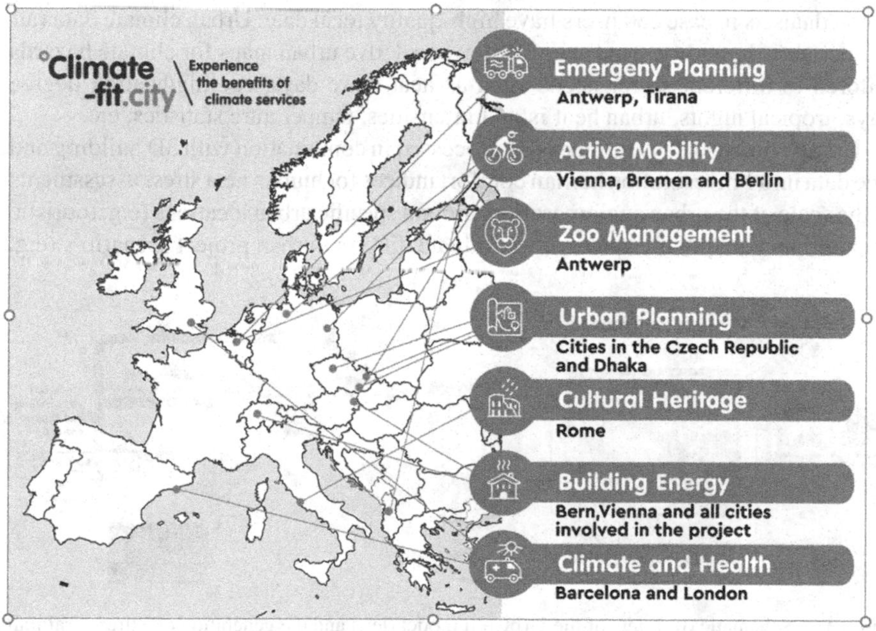

Fig. 13.1 Overview of Climate-fit.City cases worked out during the H2020 Climate-fit.City project

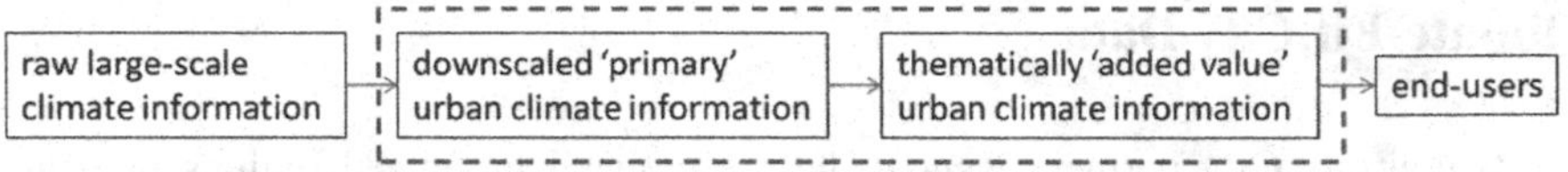

Fig. 13.2 Data flow inside Climate-fit.City. Large-scale climate information is taken from the Copernicus Climate Change Service (C3S)

- quantify socio-economic impacts of adaptation and policy support.

Climate-fit.City services are based on downscaled large-scale climate information at the city scale which is used by sectorial experts to create added value urban climate information (Fig. 13.2).

Primary urban climate data is delivered by VITO and KU Leuven. VITO focuses with its urban climate model UrbClim (De Ridder et al. 2015) on heat stress and related meteorological variables and indicators, while KU Leuven applies its rainfall downscaling statistics methodology to focus on precipitation and flooding.

The urban boundary layer climate model UrbClim is designed to cover individual cities and their nearby surroundings at a very high spatial resolution (De Ridder et al. 2015). UrbClim consists of a land surface scheme, coupled to a three-dimensional (3D) atmospheric boundary layer module. A schematic representation of the UrbClim model is presented in Fig. 13.3. For all simulations, the model is set up with a spatial resolution of 100 m. For the historical (reference) simulations, the model is driven with meteorological data from the Copernicus Climate Change Service. Model configuration is based on Copernicus land cover data combined with local urban datasets in case end users have high-quality local data. Urban climate data (air temperature, humidity, wind speed) allows to derive urban maps for climate hazards tailored to different sector needs such as heat wave days, cooling/heating degree days, tropical nights, urban heat island intensities, temperature statistics, etc.

Urban climate data is further post-processed in combination with 3D building and tree data into 1 m resolution human comfort indices for human heat stress assessments at the scale of the urban agglomeration, around specific urban locations (e.g. touristic sites, urban zoo, public squares), to evaluate different urban project scenario's (e.g.

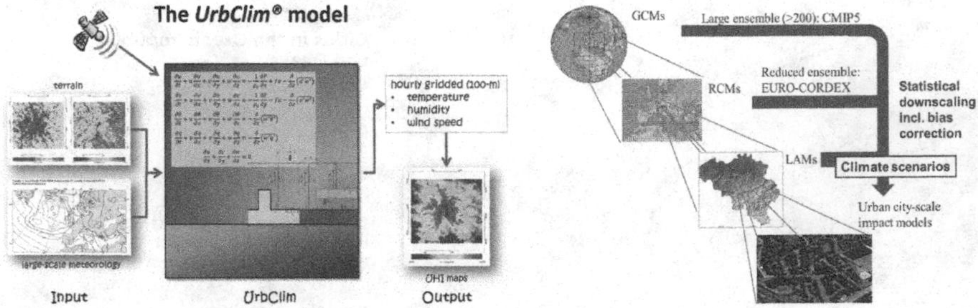

Fig. 13.3 Schematic overview of the UrbClim model (left) and the generation of extreme rainfall maps and climate scenarios (right)

greening) or to estimate the impact of heat stress on outdoor labour productivity (Lauwaet et al. 2020).

The generation of extreme rainfall maps and climate scenarios is based on a method of statistical downscaling (Willems and Vrac 2011; Willems et al. 2012) where all publicly available global and regional climate model outputs are considered and downscaled to the local scale of the city (Fig. 13.3). Local time series of measured precipitation intensity available for specific locations in Europe (i.e. cities) are perturbed according to climate change signals obtained from the climate model outputs (Hosseinzadehtalaei et al. 2017, 2018, 2020).

More details about the generation of the current and future primary urban climate data are documented in the Climate-fit.City deliverable 5.2 (available in the resource section of the project website https://www.project.climate-fit.city/). All data generated for the case study cities is freely available for download on https://www.datapl atform.climate-fit.city/.

Climate-Fit.City Services

Human activities are heavily impacted by thermal stress. During extreme heat waves, there is an uneven amplification of thermal discomfort inside the cities that leads to even higher impacts, increasing human mortality. Moreover, global climate projections consistently point towards an increase of the number, duration and intensity of heat waves (Vogel et al. 2017). Extremely hot summers such as in 2003 in Europe are likely to become fairly common towards the end of the century. Extreme heat causes direct impacts on human health due to increased mortality, discomfort and mental illnesses but also indirectly through the advance of tropical disease vectors and the deterioration of air quality. Moreover, it also reduces outdoor and indoor labour productivity, increases energy consumption for air conditioning (for human comfort but also to avoid, e.g. server and other IT infrastructure overheating), deteriorates transport infrastructure (e.g. rail buckling, road asphalt warps and melts) and impact eco-systems. Finally, it leads to lower agriculture production, makes cooling of energy production more difficult and even reduces airfreight cargo (weight limit at take-off is reduced in case of very warm air temperatures). Within Climate-fit.City, demonstrations have been worked out on the impact of heat for active mobility (focus on cycling), human health (heat–health mortality), building energy demand/indoor comfort, urban spatial planning and urban Zoo management.

Besides the heat related services, Climate-fit.City also demonstrated the use of urban pluvial flooding risk maps for climate change resilient emergency management. Finally, all services are co-designed with urban end users. An overview is given in Table 13.1.

Table 13.1 Climate-fit.City services

Name	Short description/URL
Active mobility service	Provides information on the sensitivity of a city's bicycle traffic volume towards variations in meteorological conditions and on a city's current and future climatic attractiveness towards cycling, including spatial and temporal variations. This information allows identifying regions or routes particularly exposed to meteorological conditions perceived as unattractive by the city's cyclists. The service aims at supporting climate-inclusive bicycle traffic planning. It is delivered in the form of (i) a detailed service report including maps, figures, tables and interpretation guidelines of the analysis results and (ii) new climatic features within the upgraded tool Bike Citizens Analytics.
Bike citizens analytics platform	Bike Citizens Analytics is a GPS data analysis tool from the Austrian private company Bike Citizens (https://www.bikecitizens.net/) that aims at supporting towns and cities in their bicycle traffic planning. With the incorporation of the Active Mobility Service in the Climate-fit.City project, the Bike Citizens Analytics platform now additionally offers the possibility for climate-related analyses, i.e. (i) the comparison of cycling intensities at different meteorological conditions and (ii) the visual inspection of spatial variations in typical climatic conditions (wet-bulb globe temperature, wind speeds) during daytime (see https://project.climate-fit.city/wp-content/uploads/2020/07/Video_BikeCitizensAnalytics.mp4 for a walk-through).
Building energy service	Building Energy service support buildings design/maintenance to be better adapted to the specific urban climate and future climates (indoor comfort) and to lower energy consumption for heating/cooling. Meteonorm software (https://meteonorm.com/) is upgraded in the Climate-fit.City project to include urban and climate change effects.
Zoo management	Online tool to support the Zoo in the climate change management of their animal collections, energy and water consumption and visitor marketing, taking detailed and state-of-the-art climate data into account. The tool is developed for a Zoo context but can be adapted for other types of locations/organizations as commercial activities, touristic sites, sport locations, … (https://kmda.climate-fit.city/).
Urban spatial planning	The Urban Planning Service focuses on the influence of the urban land use structure on the level and spatial distribution of heat stress in urban areas. Through modification of the input–land use layer, various city development scenarios can be simulated and the corresponding distributions of the heat stress levels in the area of interest are modelled/re-calculated. This scenario modelling service is provided on two different spatial scales: (i) city level—in 100-m spatial resolution and (ii) local level—in 1-m spatial resolution. For the city level of modelling, an interactive scenario modelling tool has been developed, enabling the user to interactively model different scenarios of the city development online and then to run directly the modification of the resulting map, showing the distribution of the heat stress levels in the city. A demonstration service was developed inside the ESA Urban-TEP platform: https://urban-tep.eu/puma/tool/

(continued)

Table 13.1 (continued)

Name	Short description/URL
Heat-health service	Spatially distributed heat mortality association data showing the increased risk of mortality for warm summer days, in comparison with tempered summer days with the influence of gender, education and age. Two demonstrator online platforms have been developed within the Climate-fit.City project: Barcelona: https://aspb.shinyapps.io/climate-fit-city-en/ London: https://londonheat.shinyapps.io/climate-fit-city/
Emergency planning service	The Emergency Planning Service delivers improved knowledge and insight on modified extreme weather conditions and related consequences, as input to set up a climate-proof city emergency plan for extreme rainfall and pluvial flood related disasters.
Socio-economic impact service	Socio-economic impact service provides an assessment of socio-economic impacts by the climate services for each demonstration case of Climate-fit.city. Different data gathering and analytical methods are employed according to the specificities of each demonstration case, the goal of the climate service and the kind of available data and stakeholders

Conclusions

The socio-economic impact assessment performed in the Climate-fit.City project found that diverse actual and potential added values exist in terms of public service effectiveness, economic impacts, policy impacts and raising awareness impacts (e.g. improve bike paths by accommodating climate needs, increase bike use and reduced carbon emissions, reduction of deaths attributable to heat waves and health cost reduction, support a revision of existing building and built environment policies, improve communication around heat-related issues, updating building policies and standards at national and local level, better allocate emergency flooding equipment, support zoo management in properly managing of energy investments, etc.).

Along the project, it has also been confirmed that the urban climate services represent a relevant tool that can provide scientific support, for example, through their maps, data and climate scenarios, that can lead to potential improvement of the effectiveness into a range of public services. On this regard, the climate services proved to potentially provide a relevant support for the development of evidence-based urban policies.

Urban climate data delivery is globally guaranteed by using satellite-based land cover, building, soil and vegetation data. However, the applicability of the sectorial service components depends on the availability of input data, for example, the heat–health service requires high-resolution mortality data. Actually, service demonstrations use separate online tools. Future potential improvements could be to integrate all components inside one service platform.

To conclude, we want to highlight the integrated nature of the Climate-fit.City service. All service components are using the same urban climate data which streamlines the application of multiple services and, secondly, the large variety of sectoral applications supports the involvement of many urban actors which is found to be a barrier in urban adaptation. Mostly, climate adaptation is seen as a responsibility of the energy, climate and environment departments which leads to reduced interest from other urban departments. Climate-fit.City provides an integrated perspective reaching out to other departments as health, mobility, urban planning and green infrastructure as well as emergency planning departments.

Acknowledgements The Climate-fit.City project was funded within the European Union's H2020 Research and Innovation Programme under Grant Agreement No. 73004. The authors also want to thank the project officer and the project reviewers for their constructive feedback.

References

De Ridder K, Lauwaet D, Maiheu B (2015) UrbClim—a fast urban boundary layer climate model. Urban Clim 12:21–48

Hosseinzadehtalaei P, Tabari H, Willems P (2018) Precipitation intensity–duration–frequency curves for central Belgium with an ensemble of EURO-CORDEX simulations, and associated uncertainties. Atmos Res 200:1–12

Hosseinzadehtalaei P, Tabari H, Willems P (2017) Uncertainty assessment for climate change impact on intense precipitation: how many model runs do we need? Int J Climatol 37(S1):1105–1117

Hosseinzadehtalaei P, Tabari H, Willems P (2020) Climate change impact on short-duration extreme precipitation and intensity-duration-frequency curves over Europe. J Hydrol 590:125249

Lauwaet D, Hooyberghs H, Maiheu B, Lefebvre W, Driesen G, Van Looy S, De Ridder K (2015) Detailed urban heat island projections for cities worldwide: dynamical downscaling CMIP5 Global climate models. Climate 3:391–415. https://doi.org/10.3390/cli3020391

Lauwaet D, Maiheu B, De Ridder K, Boënne W, Hooyberghs H, Demuzere M, Verdonck ML (2020) A new method to assess fine-scale outdoor thermal comfort for urban agglomerations. Climate 8(6). https://doi.org/10.3390/cli8010006

Vogel MM, Orth R, Cheruy F, Hagemann S, Lorenz R, Hurk BJJM, Seneviratne SI (2017) Regional amplification of projected changes in extreme temperatures strongly controlled by soil moisture temperature feedbacks. Geophys Res Lett 44:1511–1519

Willems P, Vrac M (2011) Statistical precipitation downscaling for small-scale hydrological impact investigations of climate change. J Hydrol 402:193–205

Willems P, Olsson J, Arnbjerg-Nielsen K, Beecham S, Pathirana A, Bülow Gregersen, I, Madsen H, Nguyen V-T-V (2012) Impacts of climate change on rainfall extremes and urban drainage. IWA Pub. 252 pp

Chapter 14
Weather and Climate Services to Support a Risk-Sharing Mechanism for Adaptation of the Agricultural Sector. A Theoretical Example for Drought-Prone Areas

María Máñez Costa and Dmitry V. Kovalevsky

Abstract Sharing the burden of adaptation is key for the agricultural sector in developing countries. For the agricultural sector in developing countries, the losses will go from 3% under 1.5 °C scenario to 7% under 2 °C scenario (Masson-Delmotte et al. 2018). This anticipated information on possible climate change-driven challenges possesses a big load in farmers management that might ex-ante stop investing because of the negative consequences of the scenarios presented. This situation could be even worse in subsistence farming system totally dependent on the yields. Crop insurances can be a good way to overcome some of the losses. In this paper, we present weather-based insurance schemes (WII), which are based on weather index objectively determined for the specific agricultural region, and therefore the individual loss assessment, which makes insurances too expensive, is not necessary. We present the results of decisions based on perfect and imperfect weather forecasts and conclude by offering insights in the difference of decision-making if a perfect forecast might be available or not and the consequences for farmers income.

Keywords Insurance · Agricultural sector · Climate services · Adaptation · SDGs

Introduction

One of the first things that comes to mind when talking about adaptation is the question of the sharing of the risks we might be facing. As the IPCC Special Report 1.5 Degrees pointed out, the increase of risks between 1.5 °C and 2 °C is quite frightening for several parts of the economy. Agriculture is considered the economic sector most vulnerable to climate change, and in many regions of the developing world, weather-

M. Máñez Costa (✉) · D. V. Kovalevsky
Climate Service Center Germany (GERICS), Helmholtz Zentrum Hereon, Fischertwiete 1, 20095 Hamburg, Germany
e-mail: maria.manez@hereon.de

D. V. Kovalevsky
e-mail: dmitrii.kovalevskii@hereon.de

C. Kondrup et al. (eds.), *Climate Adaptation Modelling*, Springer Climate,
https://doi.org/10.1007/978-3-030-86211-4_14

and climate-related risks for the agricultural sector are already jeopardizing parts of the economy dependent on the agricultural outputs (Guimarães Nobre et al. 2019). Since the 1980s, climatic risks have diminished worldwide the yields of main staple food (e.g. maize by 3.8% and wheat 5.5%) (Lobell et al. 2011). The resulting impacts are not only for farmers but also for all the upstream and downstream sectors of those countries.

To share the burden of this risk is key for reaching the Sustainable Development Goals (SDGs), mostly SDG2 related to the reduction of hunger to zero. This goal is crucial for around 700 Million people that are reported to go to bed hungry every day (FAO et al. 2020).

Additional to the SDGs, the Sendai Framework and the Paris Agreement also identify risk-sharing as an important mechanism for facing climate-related risks. Through risk transfer and insurance, risk-sharing turns fundamental to enhance resilience and decrease the economic impacts of climate-related risks.

A crop insurance is one of the possible risks sharing mechanisms that might be implemented to share the burden and be able to adapt to climate-related challenges. Crop insurance is an important adaptation mechanism for agricultural risks because it redistributes the burden of the risk between the farmers and the insurance companies. But, in general, it is often expensive and unaffordable to many farmers, especially in low-income countries. One of the reasons for why crop insurance premiums are so high is the expenditures for loss assessment.

A possible crop insurance mechanism is given by weather-based index insurances. In weather-based index insurance (WII) schemes, the payouts are based on weather index objectively determined for the specific agricultural region, and therefore the individual loss assessment is not necessary. WII are designed not for the individual risk but more for the climate risks that might influence a region's resilience. Climate services designed for WII are key.

WII paves the way to cheaper and more affordable agricultural insurances (Jørgensen et al. 2020). We suggest that redistributing the risk of agricultural failures through crop WII insurances is an appropriate adaptation strategy that also might support SDG 2 because the implementation of weather-based insurances indicates a high risk-reducing potential, therefore it increases the resilience of farming systems towards risks. Additionally, broad implementation of weather insurance schemes would support not only SDG 2 on zero hunger, but also SDG 10 on reducing inequalities and SDG 13 on climate action (UN 2019).

Weather-Based Index Insurance (WII)

Climate-related extremes as presented in the IPCC 1,5 Special report, can have not only ex-post consequences after, e.g. heavy rains have caused flooding, but also ex-ante consequences because knowing the risk might dissuade from investing in the agricultural sector. In both situations, the losses are high, on the one side because of the damages provoked by flooding and on the other side because of the low productivity due to insufficient investment. Therefore, adaptation instruments should support farming systems in both mentioned situations, by supporting with the sharing of the risk burden and by enabling investments in the agricultural sector.

WII are effective adaptation instruments for ex-ante and ex-post climate risk consequences. They are based on measurable climate variables as precipitation causing, for example, heavy rains or temperature causing heat waves. In WII, losses and payout are determined by the measurement in a particular region. Insurance takers can buy insurances at a relative low cost and this way re-distribute the risk.

WII aggregates the weather information within the vegetation period. This might be a rather simple index, e.g. based on precipitation only, or a more complex construct additionally dependent on, e.g. daily air temperature through thermal units or Growing Degree Days (GDD) which is a measure of heat accumulation that parametrizes the heat affecting the plant (Kovalevsky and Máñez Costa 2021).

One important question when deciding on buying WII is the existence or not of proper weather and climate services that could be used to decide for farmer A, if a WII is worth to buy or not. In many cases, the fact is that not all over the developing world, we will be able to get enough data or access to data that will allow farmers to have the science-based information support. Therefore, our main question to answer was: under which conditions farmers in drought-prone areas will buy a WII and how will this affect their income. With our modelling approach we answer these questions.

Overview of Modelling Approach

We will show that developing comprehensive models of WII schemes is important as adaptation instruments, as WII could have a great impact on making the access to insurances easier for farmers than it is today. As already mentioned, nowadays agricultural insurances are only affordable to rich farmers. This kind of insurance would also support the capability of farmers to have a back-up monetary solution in case of crop failure (ex-post). Additionally, we assume that it would be easier to enhance the resilience of farming systems through WII when WII projects are supported by reliable forecasting tools.

The presented study contributes to theoretical modelling of certain dynamical features of WII project implementation. Specifically, it is focused on modelling the strategies of producers in a drought-prone region regarding the WII policy purchase, and also on simulating the dynamics of aggregate demand for this kind of insurance (see Fig. 14.1). Strategies of individual producers might depend on availability of the weather index forecast and on its quality. The analysis performed suggests that the quality of the forecast would affect the optimal strategy to be selected by an insurance policy buyer under conditions of inevitable uncertainty. Modelling approaches developed can support decisions relevant to design, successful implementation and subsequent scale-up of WII schemes in regions prone to agricultural droughts.

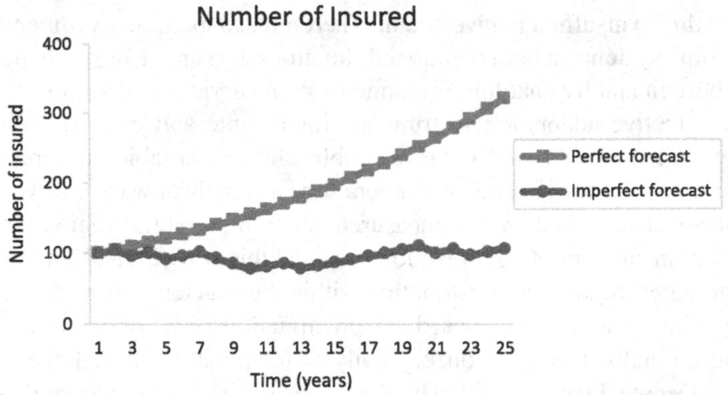

Fig. 14.1 Simulated demand for WII depending on perfect and imperfect forecast for weather index. In case of a perfect forecast (one that can be proven by looking at the previous time period forecasted), there is a steady growth in number of insured farmers. In case of an imperfect forecast, the numbers might be very stable or even decreasing

The analysis of adverse selection is made for a hypothetical scenario where only the policy buyers, but not the insurers, base their decision-making on available weather index forecast. An analogous scenario was considered in a study devoted to rainfall index insurance in the US, together with a more realistic scenario where both the insurers and the insured take into account weather index forecast (Nadolnyak and Vedenov 2013). As shown in Nadolnyak and Vedenov (2013), in the latter case, adverse selection can be precluded by setting insurance premiums on the basis of forecast-conditional calculations. The generalization of the model developed in the present study to account for the decision-making of insurers is left for further research.

Our findings are based on a simple conceptual model for WII. In particular, a simple power law was chosen to parameterize the dependence of crop yield on the weather index. Also, the dynamics of both actual and forecasted weather index was, somewhat arbitrarily, represented by a very simple random process. Bringing more realism to the model, in particular, adopting the crop yield parametrizations from real-world field data, and deriving actual/forecasted weather indices from observations/forecasts of weather and climate variables, is planned for future research. Implementation of this program would make the proposed generic modelling scheme crop- and region-specific, which, as mentioned above, is a necessary prerequisite for successful real-world implementation of WII projects. Another interesting extension of this modelling scheme would be to consider the case where probabilistic forecasts are available to the insurers and the insured, providing not only the expected values of the weather index, but additionally their confidence intervals.

Conclusions and Recommendations

The dynamics of individual producer's income are shown in Fig. 14.2 (blue line: 'always insurance' case; green line: hypothetical perfect forecast case; red line: imperfect forecast case). Producers supported in their decision-making by hypothetical perfect forecasts would benefit from 'perfectly efficient' adverse selection. This makes the dynamics of producers' income under a perfect forecast scenario quite different from the imperfect forecast case (where the forecast quality is low and hence the probability of a wrong decision is high) and from the 'always insurance' case (where, by assumption, producers bear expenditures for insurance policies in all years, 'bad' and 'good').

In a simple WII model discussed in the present paper, the stochastic dynamics of actual weather index were simulated with a stationary random process. However, the climate is changing. In particular, climate model projections suggest that anthropogenic global warming will in future lead to many wet areas getting wetter and to dry regions getting drier. Generally, model projections reveal many drying areas in the low- and mid-latitudes, with a tendency for the subtropical (dry) zones to expand poleward. Consequently, future droughts will have more negative impacts on many economic activities, especially on agriculture, which is, as mentioned at the beginning, the most sensitive sector. By generalizing the proposed modelling scheme to the case of non-stationary random processes, it would be possible to consider weather indices with statistically significant trends that would be important for taking climate change effects into consideration. This would also likely shift the prospects for applications of this modelling scheme to the rapidly developing area of climate services. This modelling also shows the differences in income depending on the decisions made by farmers on buying or not an insurance policy.

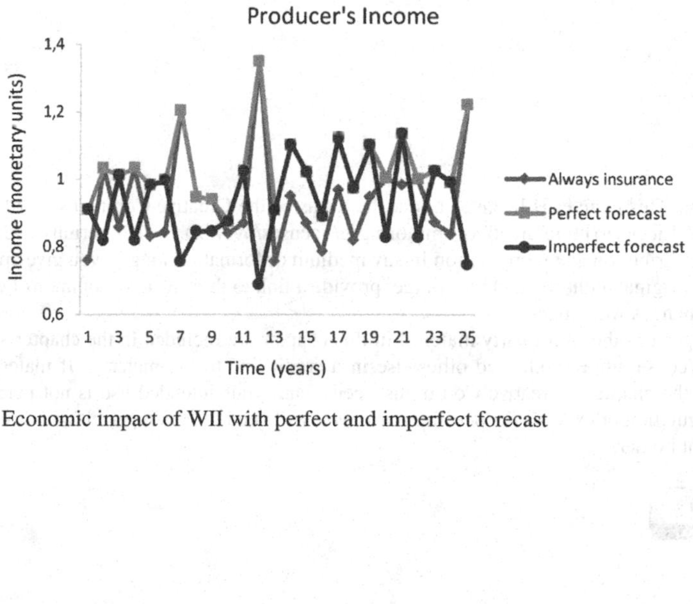

Fig. 14.2 Economic impact of WII with perfect and imperfect forecast

References

FAO et al (2020) The State of Food Security and Nutrition in the World (SOFI). Rome, Italy

Jørgensen, Liv S, Termansen M, Pascual U (2020) Natural insurance as condition for market insurance: climate change adaptation in agriculture. Ecol Econs 169:106489. http://www.sciencedirect.com/science/article/pii/S0921800919301612

Kovalevsky DV, Máñez Costa M (2021) Demand behaviour for weather index insurance products in regions prone to agricultural droughts. Discontinuity, Nonlinearity, and Complexity 10(4):765–780. DOI: https://doi.org/10.5890/DNC.2021.12.015 (open access)

Lobell DB, Schlenker W, Costa-Roberts J (2011) Climate trends and global crop production since 1980. Science 333(6042):616–620

Masson-Delmotte V, Zhai P, Pörtner H-O, Roberts D, Skea J, Shukla PR, Pirani A, Moufouma-Okia, Péan C, Pidcock R, Connors S, Matthews JBR, Chen Y, Zhou X, Gomis MI, Lonnoy E, Maycock, Tignor M, Waterfield T (eds) (2018) Global warming of 1.5 °C. An IPCC Special Report on the impacts of global warming of 1.5 °C above pre-industrial levels and related global greenhouse gas emission pathways, in the context of strengthening the global response to the threat of climate change. Geneve

Nadolnyak DA, Vedenov D (2013) Information value of climate forecasts for rainfall index insurance for Pasture, Rangeland, and Forage in the Southeast United States. J Agri Appl Econ 45(1):143638. https://econpapers.repec.org/RePEc:ags:joaaec:143638

Nobre G, Gabriela et al (2019) Financing agricultural drought risk through Ex-Ante cash transfers. Sci Total Environ 653:523–535. http://www.sciencedirect.com/science/article/pii/S0048969718343067

UN (2019) The Sustainable Development Goals Report. New York

Chapter 15
Recent Innovations in Flood Hazard Modelling Over Large Data Sparse Regions

Jeffrey Neal

Abstract This opinion piece summarises recent progress in the development of global flood models (GFMs) in support of flood impact modelling and identifies potential areas for model improvement over the next 5–10 years. In many parts of the world, flood hazard data are absent or lack the accuracy and precision required for most practical applications, including climate change impact assessment. With the hydrological cycle expected to intensify due to climate change, better modelling of flood hazard is needed as a prerequisite to understanding how flood risk might change in the future with climate. The past decade has seen substantial advances in the modelling of flood inundation in data scarce areas along with the emergence of global flood models that could form the foundation of global impact assessments. In summarising these advances, four key themes emerge linked to topography, extreme flow estimation, river network parameterisation and numerical modelling of inundation. Progress in each of these themes will be needed to deliver the next generation of global flood hazard data.

Keywords Flooding · Flood risk · Flood hazard · Global flood modelling

Introduction

The mapping of flood hazards and risk via numerical modelling has become an integral component of flood risk management in many advanced economies. The benefits of these data with respect to land planning, insurance provision and disaster response are well established. Yet, for much of the world, flood hazard data are absent or lack the accuracy and precision required for most practical applications, including climate change impact assessment. With the hydrological cycle expected to intensify due to climate change, more accurate modelling of where is at risk from flooding is required as a prerequisite to understanding how flood risk might change with the

---

J. Neal (✉)
School of Geographical Sciences, University of Bristol, Bristol BS8 1SS, UK
e-mail: j.neal@fathom.global; j.neal@bristol.ac.uk

Square Works, 17-18 Berkeley Square, Clifton, Bristol BS8 1HB, UK

C. Kondrup et al. (eds.), *Climate Adaptation Modelling*, Springer Climate,
https://doi.org/10.1007/978-3-030-86211-4_15

climate. The past decade has seen rapid advances in the modelling of flood hazards in data scarce areas where the traditional local scale engineering approaches used in developed nations are not possible. Since it tends to be possible to automate the production of such models, a focus on continental to global scales applications has emerged. An example of the output from a GFM can be seen in Fig. 15.1 where flood depth is plotted for the region surrounding Bangkok for a 1 in 100-year flood hazard. Here, recent innovations in the field of global flood hazard mapping are reviewed with a steer as to how these might support enhanced climate impact assessment.

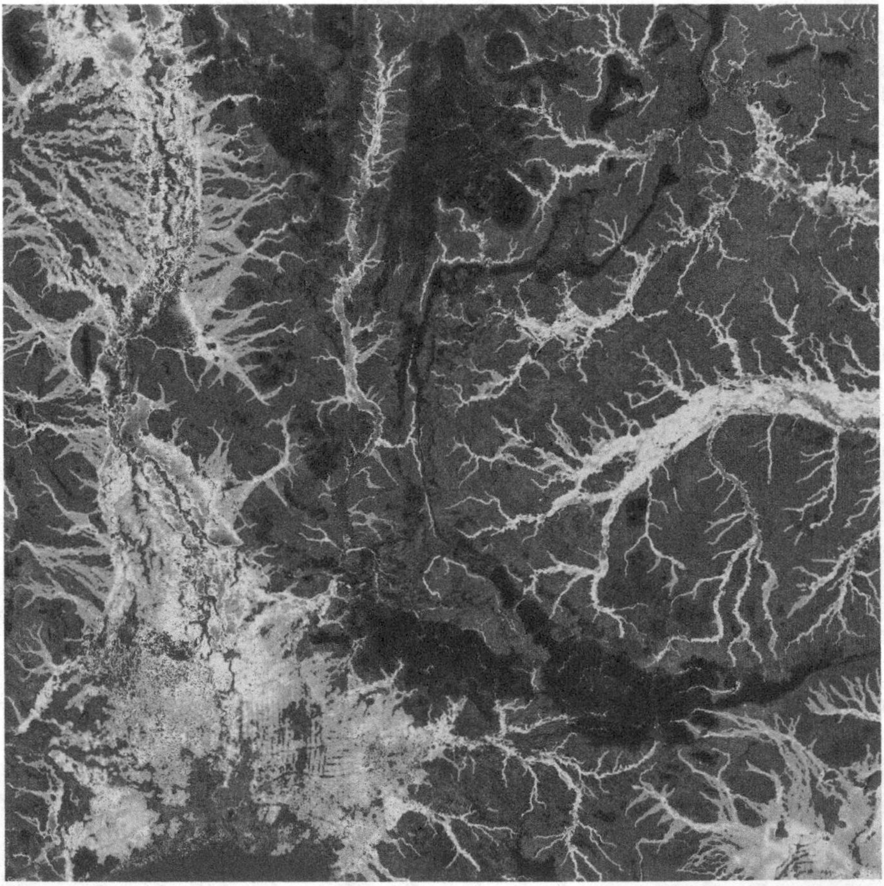

Fig. 15.1 Flood inundation depth for the 1 in 100-year flood hazard, simulated by the Fathom GFM (Sampson et al. 2015)

Overview of Progress in Global Flood Modelling

The ability of a GFM to estimate flood hazard is broadly contingent on four components, these are:

The model of terrain elevations (DEM).
The method used to estimate extreme flows.
The definition of the river network.
The numerical model to simulate inundation.

Below, recent advances in each of these components are summarised, sometimes with a focus on the GFM developed by the University of Bristol to keep the discussion brief. Ongoing data needs and key modelling uncertainties are identified along with some opportunities to improve the models over the next 5–10 years. An opinion on the current state of the field is then provided at the end.

Digital Terrain Modelling

Floods are shallow waves with long wavelength and low amplitudes. As such, they are highly sensitive to the terrain over which they flow, which can both alter and block flow pathways. It is widely accepted that airborne LiDAR data offer the most accurate terrain data for flood mapping, with sub-metre resolution and vertical errors in the low decimetres. However, LiDAR data are absent in data sparse areas and global scale DEM data must be used instead. For much of the last two decades, data obtained by the Shuttle Radar Topography Mission (SRTM) has been the preeminent terrain data source for flood inundation mapping in data sparse regions, and the latest revisions to these data seek to remove a multiple sources of vertical errors including stripe noise, random errors, absolute bias, vegetation bias and urban biases due to buildings (Yamazaki et al. 2017). The impact of these error removal processes on the DEM can be substantial, as seen for the example from the Mekong Delta in Fig. 15.2. These data should be superseded by more accurate elevation models in the near future. For example, the TanDEM-X DEM at 90 m can in theory support more accurate flood simulation than SRTM-based DEMs (Hawker et al. 2019). However, the TanDEM-X DEM has yet to have vegetation biases systematically removed from the open data products, inhibiting its uptake by GFMs. Further advances in terrain data are most likely to come from very higher resolution proprietary datasets such as satellite photogrammetry (<2 m) and the 12.5 m version of the TanDEM-X DEM. An increased availability of such data at reduced costs is essential if global terrain data are to drive substantial improvements in global flood hazard modelling in the near future.

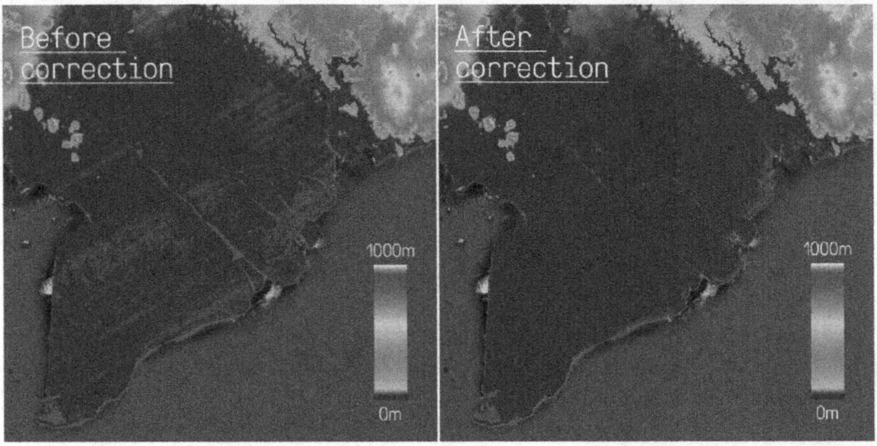

Fig. 15.2 Difference between the SRTM DEM (left) and MERIT DEM (right) over the Mekong Delta. Note the wavy stripe noise in the Mekong that means the delta elevations appear to undulate from the north west to south east in the SRTM data

Extreme Flows

For extreme event simulation, it is necessary to estimate flows at ungauged sites using either regionalisation of extreme discharge observations from gauging stations or discharges simulated by a hydrological model over a long period. The gauged approaches (e.g. Smith et al. (2015)) benefit from regionalising direct observations of extreme flows, which simplifies the modelling process and can easily take advantage of new data sets and machine learning methods. They are however limited by data scarcity in many parts of the world, short record lengths and trends in river flows that mean the time series may not be representative of the present-day conditions. Although, the observational evidence that climate change has an impact on extreme river flows is weak, with discharge trends rarely clear outside of catchments that have experienced substantial human modification. Considering climate impacts with this method is therefore difficult and many studies look at the sensitivity of hazard and risk to event magnitude rather than climate change. Estimating extreme flows from hydrological modelling is appealing because flow estimates can be made for any location and the models can be forced by either observed or simulated weather (Alfieri et al. 2017). However, substantial uncertainties in the forcing, model structure and parameterisation of large-scale hydrological models means that biases can be expected in flow simulation along with regional differences in model performance. Climate impact studies often take this approach because the modelling cascade includes variables of direct relevance to climate (e.g. precipitation, temperature). Nevertheless, a detailed intercomparison of GFMs based on the gauged and hydrological modelling methods have yet to be undertaken and numerous advancements in local scale modelling have yet to be applied at global scales. Thus, further work is needed to understand the value of each approach and the potential for multi-model ensemble prediction.

Hydrography: River Location, Width and Depth

Open water is relatively simple to observe from satellite platforms, yet only recently have comprehensive global data sets on river width and location been developed. Prior to these studies, and still for many GFMs, the location and size of rivers was based on digital terrain data, with the HydroSHEDS https://hydrosheds.org/ data sets by far the most widely used. For steep catchments, this approach can be highly effective; however, in areas of low relief (e.g. deltas), the mapping of river locations based on topography often places rivers in the wrong location. River bifurcations and human alterations are also absent from such data sets. River networks that merge terrain derived rivers, map data and surface water observations have only recently begun to emerge, but should enable substantial improvements to GFM hydrography (Yamazaki et al. 2019). Perhaps a more fundamental issue for GFMs is the parameterisation of river conveyance capacity and specifically river depth and friction. These are not observable from satellite platforms and the conveyance capacity of rivers has been extensively altered via levy construction and channel modification, for which data are often poor or not openly available. Most GFMs make an assumption regarding the conveyance capacity of the river system linked to discharge return period, which conveniently acts as a form of bias correction for magnitude errors in the extreme flow generation process (Sampson et al. 2015). Inversion of river bathymetry from surface water dynamics perhaps offers the greatest potential for a paradigm shift in GFM hydrography, with the upcoming surface water and ocean topography satellite mission providing the necessary data for the world's larger rivers.

Inundation Modelling

Numerical modelling of floodplain inundation has a substantial development history at the reach scale using computationally expensive hydrodynamic models based on shallow water flow theory. However, early GFMs tended to be extensions to the simpler river routing models used for global hydrological and land surface modelling, which estimate inundation by computing a volume excess given river channel conveyance and distributing this volume across the lowest points in the DEM (Winsemius et al. 2013). These methods are simple to implement, however, the simulations are usually less accurate than those from hydrodynamic modelling approaches. The development of more efficient hydrodynamic models and ongoing reductions in computing costs have enabled global scale hydrodynamic models to emerge (Sampson et al. 2015). Initially, these models were developed at relatively coarse resolutions for inundation simulation ($>=1$ km), however, recent models have simulated inundation over two-dimensional grids at resolutions down to 30 m. These improvements to both process representation and resolution have had substantial impacts on the estimation of flood exposure over large scales because resolution and process inaccuracy tend to bias simulations towards greater exposure estimates. This occurs mainly for two reasons. Firstly, the flat nature of floodplains means it is easy to fill a floodplain to the surrounding topography with a simple volume excess model, which then has little sensitivity to event magnitude. Secondly, people tend to live and place assets adjacent to, but not on, floodplains. Thus, any loss of resolution in the hazard or exposure data sets tends to unintentionally capture these objects within the inundated floodplain (Smith et al. 2019).

Discussion

A substantial challenge associated with global flood hazard simulation is that all the components listed above are needed to estimate hazard and the necessary sophistication of each is contingent on the others. For example, it is only worth using a more accurate, yet computationally expensive, numerical scheme if the definition of the river network puts the river in the correct place. Furthermore, since each component has been advancing rapidly over the past decade, every global flood model has a different mix of component parts to the extent that understanding model uncertainties and benchmarking models has been near impossible to date. It is also possible for a seemingly sophisticated GFM to be let down by one of its component parts, for example, if vegetation and speckle noise has not been removed from the DEM, a complex two-dimensional hydrodynamic model is unlikely to outperform a simpler method because important flow pathways will be blocked. Only limited intercomparison of GFMs has been possible to date, but the few studies to be completed have identified substantial differences between GFMs, to the extent that they disagree on where is at risk more often than they agree (Trigg et al. 2016). Validation studies on individual GFMs usually conclude they are more accurate than the model benchmarking suggests, indicating that the validation studies to date are far from comprehensive, must pick easy to simulate locations and that the accuracy of GFMs is highly variable.

References

Alfieri L, Bisselink B, Dottori F, Naumann G, de Roo A, Salamon P et al (2017) Global projections of river flood risk in a warmer world. Earths Future 5(2):171–182

Hawker L, Neal J, Bates P (2019) Accuracy assessment of the TanDEM-X 90 digital elevation model for selected floodplain sites. Remote Sens Environ 232:111319

Sampson CC, Smith AM, Bates PB, Neal JC, Alfieri L, Freer JE (2015) A high-resolution global flood hazard model. Water Resour Res 51(9):7358–7381

Smith A, Bates PD, Wing O, Sampson C, Quinn N, Neal J (2019) New estimates of flood exposure in developing countries using high-resolution population data. Nat Commun 10(1):1814

Smith A, Sampson C, Bates P (2015) Regional flood frequency analysis at the global scale. Water Resour Res 51(1):539–553

Trigg M, Birch C, Neal J, Bates P, Smith A, Sampson C et al (2016) The credibility challenge for global fluvial flood risk analysis. Environ Res Lett 11(9)

Winsemius HC, Van Beek LPH, Jongman B, Ward PJ, Bouwman A (2013) A framework for global river flood risk assessments. Hydrol Earth Syst Sci 17(5):1871–1892

Yamazaki D, Ikeshima D, Sosa J, Bates PD, Allen GH, Pavelsky TM (2019) MERIT Hydro: a high-resolution global hydrography map based on latest topography dataset. Water Resour Res 55(6):5053–5073

Yamazaki D, Ikeshima D, Tawatari R, Yamaguchi T, O'Loughlin F, Neal JC et al (2017) A high-accuracy map of global terrain elevations. Geophys Res Lett 44(11):5844–5853

Part III
Sectoral Models for Impact and Adaptation Assessment

Jaroslav Mysiak
Fondazione CMCC, Centro Euro-Mediterraneo Sui Cambiamenti Climatici, Lecce, Italy
jaroslav.mysiak@cmcc.it

Introduction

This section describes a variety of models addressing the impacts of climate change and performance of adaptation measures on various economic sectors, productive assets or services, business operation and continuity and the environment

Maire et al. describes a framework which couples and unifies climate, vegetation, land management and commodity trade models. Climate emulator simulates effects of climate change which are used by dynamic vegetation models to assess the impacts on crop yields. A socio-economic land-use model simulates demand and trade of agricultural commodities, while adjusting commodity prices rather than assuming market equilibrium conditions. The estimated short-term surplus and deficits along with environmental constraints drive adaptation responses in land management. Greenhouse gas emissions from land-use change and land management feedback to simulated climate. The socio-economic model is also able to capture the relationship between food demand, income levels and food prices. The resulting modelling framework represents trade-offs, responses and cross-scale and international interactions within a dynamic system.

Borgomeo reviews advances in decision analysis and simulation used to inform climate adaptation in water sector. Water Resource System models are forced by climatic boundary conditions to simulate river and groundwater discharge. System simulation models represent rules for water withdrawal, storage and allocation of water across various users. Water use models project water demands from households, agriculture and other economic sectors. Water resource system simulators test the performance of the system under changing conditions of climate and demand.

Pulido-Velazquez et al. review the progress made in advances describe a combination of top-down and bottom-up approaches to assess climate change impacts at the local scale with vulnerability assessment and definition of socioeconomic scenarios and adaptation options through participative methods (BU approach). The combined approach provides a systematic and practical method for supporting the selection of adaptation measures at the basin by comprehensively integrating the goals of economic efficiency.

Pant reviews challenges and advancements in modelling vulnerability of large-scale infrastructure networks. Impacts of weather and climate risk on interdependent infrastructure networks are complex, non-linear and often propagating across a part or the entire network. Contemporary climate risk models do not take into account adequately the sensitivities of cascading failure mechanisms across infrastructure networks. Better models need to take into account the probabilities, often correlated, and the intensities of weather- and climate-related extremes, networks' exposure and vulnerability to these events, network connectivity through which the cascading effects are propagated and effects of networks' disruption on society and economy. Climate proofing of infrastructure systems includes revising design standards, incorporating backup options to substitute for disrupted services, increasing network redundancy, rerouting options and speeding up the recovery of damaged assets.

Pérez-Blanco describes an interdisciplinary, replicable and scalable framework which combines methods developed in (socio-)hydrology and economics to assess implications of adaptation policies in complex human-water systems and control trade-offs between robustness of adaptation choices and their performance. Whereas microeconomic models can represent the behaviour of individuals or firms and macroeconomic models can replicate interrelations among sectors, hydrologic models study the movement, distribution and quality of water. Despite the recent progress made, conventional hydroeconomic models are not able to capture reciprocal feedbacks between human and water systems. *Socio-hydrological* models help to overcome this gap by designing modular architectures that connect the underlying economic and hydrological models through protocols describing the rules governing the exchange of information between the independent modules.

Sanderson and Stridsland address transitional risks as defined by the Task Force on Climate-Related Risk Disclosure (TCFD) and which include risks associated with market and technology shifts, loss of reputation and policy and legal challenges. While many companies assess and report the greenhouse gas (GHG) emission using GHG Protocol, they concentrate on emissions resulting from their own operations and emissions associated with consumption of energy or water. The emissions from upstream and downstream activities are often neglected. For the latter can be addressed by Environmentally Extended Input Output models, Life Cycle Assessments or a combination of both (hybrid models). The authors describe the advantages and disadvantages of the various model types and argue that only the entire value chain related emission analysis and methods foster the much needed transparency of emissions accounting.

Insurance and risk transfer instruments can not only be used to provide protection against climate-related damage and losses, they can be designed to reward individual or collective risk reduction and adaptation to climate change. Scholer and Schuermans offer several examples in which insurance can play an important role in driving risk-infomed choices. First, discounts on insurance premiums and more favourable contractual terms and conditions are incentives for policyholders to protect their property against damage caused by extreme weather and climate conditions. Second, insurers can contribute to improve the understanding of the risks associated with business interruptions, which in turn can stimulate a more pro-active management of risks. Often, no or only simplified assessment of business interruption risk is conducted, and coverage of the underlying losses is only an add-on to property insurance. Third, mortgage insurance and trade credit insurance, if designed for this purpose, can stimulate risk reduction from climate change-related risks.

A large body of research and policy literature converges that there is a need for a deep, radical or fundamental change, as opposed to marginal or incremental change, in the way development is conceived and practiced and global environmental change understood and coped with. Murray and Chadborn draw a parallel between behavioural changes endorsed as a response to the Covid-19 pandemics and those critically important for adaptation to climate change. They argue that successful transformations build upon an understanding of what behaviour is involved in driving change and what combination of behavioural intervention works best to achieve the intended change. The Interventions Ladder of Bioethics and Behaviour Change Wheel frameworks can help designing the intervention logic and categorising the instruments such as legislation, regulations, fiscal measures, guidelines, service provision, communications, media networks and marketing, environmental/social planning.

Chapter 16
A New Modelling Approach to Adaptation-Mitigation in the Land System

Juliette Maire, Peter Alexander, Peter Anthoni, Chris Huntingford, Thomas A. M. Pugh, Sam Rabin, Mark Rounsevell, and Almut Arneth

Abstract Climate change, growing populations and economic shocks are adding pressure on the global agricultural system's ability to feed the world. In addition to curbing the emissions from fossil fuel use, land-based actions are seen as essential in the effort to mitigate climate change, but these tend to reduce areas available for food production, thereby further increasing this pressure. The actors of the food system have the capacity to respond and adapt to changes in climate, and thereby reduce the negative consequences, while potentially creating additional challenges, including further greenhouse gas emissions. The food system actors may respond autonomously based on economic drivers and other factors to adapt to climate change, whereas policy measures are usually needed for mitigation actions to be implemented. Much research and policy focus has been given to land-based climate change mitigation, but far less emphasis has to date been given to the understanding of adaptation, or the interaction between adaptation and mitigation in the land use and food system. Here, we present an approach to better understand and plan these interactions through modelling. Climate change adaptation and mitigation strategies and the impacts on

---

J. Maire · P. Alexander (✉) · M. Rounsevell
School of Geosciences, University of Edinburgh, Edinburgh, UK
e-mail: peter.alexander@ed.ac.uk

P. Alexander
Global Academy of Agriculture and Food Security, The Royal (Dick) School of Veterinary Studies, University of Edinburgh, Edinburgh, UK

P. Anthoni · S. Rabin · M. Rounsevell · A. Arneth (✉)
Institute of Meteorology and Climate Research/Atmospheric Environmental Research, Karlsruhe Institute of Technology, Garmisch-Partenkirchen, Germany
e-mail: almut.arneth@kit.edu

C. Huntingford
UK Centre for Ecology and Hydrology, Wallingford, United Kingdom

T. A. M. Pugh
Birmingham Institute of Forest Research, University of Birmingham, Birmingham, UK

School of Geography, Earth and Environmental Sciences, University of Birmingham, Birmingham, UK

C. Kondrup et al. (eds.), *Climate Adaptation Modelling*, Springer Climate,
https://doi.org/10.1007/978-3-030-86211-4_16

the global food system and socio-economic development can be simulated over long-term predictions, thanks to the new combination of multiple models into the Land System Modular Model (LandSyMM). LandSyMM takes into account the impacts in changes in climate (i.e. temperature, precipitation, atmospheric greenhouse gas concentrations) and land management on crop yields with its implications for land allocation, food security and trade. This new coupled model integrates, over fine spatial scale, the interactions between commodities consumption, land use management, vegetation and climate into a worldwide dynamic economic system. This study offers an outline description of the LandSyMM as well as the perspectives of uses for climate adaptation assessment.

Keywords Land-use change · Dynamic global vegetation model · Climate change · Food system

Introduction

Food production systems are interlinked with the efforts to tackle climate change impacts. Currently, the food production system accounts for about one-third of the global greenhouse gas emissions, and 50% of the global habitable land (ice- and desert-free) is used for agriculture (IPCC 2019). Land use and land-use change are associated with a quarter of global greenhouse gas emissions from mainly tropical deforestation, methane emissions from livestock and rice cultivation, nitrous oxide emission from fertilized soils and manure management (IPCC 2019). Therefore, land use changes are contributing greatly to climate change as 11% of anthropogenic carbon dioxide (CO_2) emissions are associated with land use change (Friedlingstein et al. 2019), but also they play a key role in adaptation to the impact of climate change on agriculture (Alexander et al. 2018; Agnolucci et al. 2020).

Adaptation to climate change intends to moderate potential damages from climate change along with benefitting from opportunities associated with climate change impacts. The global land use and food system has the capacity to respond and adapt to changes to the climate, and thereby reduce some of the negative consequences of these changes, while potentially creating additional challenges. Adaptive mechanisms are both direct, i.e. as a response to climatic conditions in that location, and indirect, e.g. in response to market movements or policy decision themselves created by environmental changes, including those that may be occurring in other locations. Examples of food production adaptations to climate change include altering agricultural practices, such as choice of crop types or intensity of management, or shifting cultivated areas within and between countries. The wider food system also has capacity to adapt, e.g. through shifts in patterns of consumption. Shifts in consumer perception and preferences (e.g. the rise in vegetarianism and veganism), as well as changes mediated by market prices are both likely to be important for these demand-driven adaptations. And these adaptations may have climate-change mitigation co-benefits through reducing greenhouse gas emissions from land use, as well as reducing fresh-water over-use, water and air pollution, protecting wildlife, restoring lands back to forests or grassland (Rabin et al. 2020).

Modelling approaches are essential to help stakeholders to develop policies toward long-term actions for better food production systems, which are more resilient to climate change impacts and at the same time, contribute to halting climate change. While some of these adaptations may be actively steered by policy, actors throughout the food system will also adjust based on economic and other factors. Part of this autonomous adaptation includes land managers and farmers making decisions that can negatively interact with policymakers' agendas; e.g. intensifying production on existing agricultural land. Land-use modelling offers a unique chance to simulate the impacts of the adoption of land-based climate change mitigation measures, the role of the different actors along the food systems, the effect of the continuing globalization of trade in food products and increasing demand for agricultural goods (Humpenöder et al. 2015). However, currently available land use based models do not focus on adaptation responses to climate change in land use, but more on a top-down mitigation policies neglecting the implementation of the small-scale actors adaptation decisions (Alexander et al. 2018; Robinson et al. 2018). Here, we describe a coupled model system, the Land System Modular Model (LandSyMM) which aims to support future climate research by assessing the interplay between natural system dynamics and socio-economic processes related to supply and demand. The multiple scales represented allow interactions of bottom-up adaptation dynamics and top-down mitigation policies to be represented and explored (Müller et al. 2020).

Modelling Adaptation in the Land Use and Food System

Existing Approaches and Research Gaps

A range of models have attempted to understand how future agricultural and land use systems will affect and be affected by climate changes. These models have highlighted key societal drivers and were applied to a wide range of scenarios (e.g. greenhouse gas emissions and radiative forcing, socio-economic pathways). However, due to computational restrictions, most of the existing models typically use a very low spatial resolution (Robinson et al. 2014). The downside of such an approach is that it cannot well account for physical limitations of productivity and does not relate to location-specific yield response to agricultural changes in inputs (Alexander et al. 2018). Moreover, current model applications tend to focus on the climate change mitigation potential of land use rather than placing adaptation at their core. Adaptation requires information at much finer spatial resolution than can be typically provided in integrated assessment models (IAMs). The resolution of the LandSyMM enables us to explore adaptation measures such as related to crop productivity variations from changes in management practices (e.g. fertilizer and irrigation rates), or management in forests.

LandSyMM Modelling Approach

LandSyMM couples a dynamic global vegetation model (LPJ-GUESS; Smith et al. 2014), a climate system emulator (IMOGEN; Huntingford et al. 2010) and a socio-economic land-use model (PLUMv2; Rounsevell et al. 2014; Engström et al. 2016; Alexander et al., 2018) (Fig. 16.1). LandSyMM is currently being run at 0.5° spatial resolution. The dynamic global vegetation model (DGVM) computes for example changes in crop yields at a given location, in response to climate change, irrigation and fertilizer application which can be adjusted flexibly as part of mitigation strategies. PLUM simulates demand and trade of commodities (e.g. cereals, oil crops, pulses, starchy roots, sugar, fruits and vegetables, wood, dairy and meat products from ruminant livestock and monogastric livestock) based on least-cost optimization principles by adjusting commodity prices instead of assuming market equilibrium, allowing short-term surplus and deficits. This includes also costs for irrigation, fertilizer use and management intensity (e.g. pesticide and machinery use). LPJ-GUESS water runoff outputs are used by PLUM to constrain irrigation use from water availability at the basin level, after adjusting for other uses and environmental limitations. Therefore, changes in water resources as well as plant requirement under future climates can drive adaptation responses in land management. PLUM captures the relationship

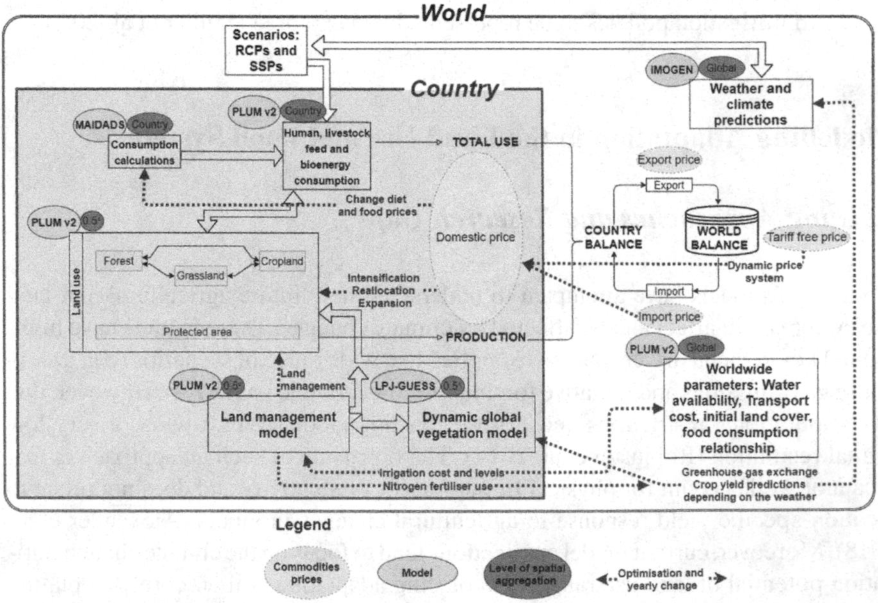

Fig. 16.1 LandSyMM structural overview. The focus of the schematic is on the cross-scale interactions between models (PLUMv2, LPJ-GUESS, IMOGEN, MAIDADS) and the embodied interactions between the country- and world-level calculations for each time step of the model. GDP: Gross Domestic Product; RCPs: representative concentration pathways; SSPs: shared socio-economic pathways

between food demand in each country and income and food prices using a Modified, Implicit, Directly Additive Demand System (MAIDADS) approach (Preckel et al. 2010; Gouel and Guimbard 2018). Prices are endogenous in PLUM and are adjusted through international commodity trade imbalances, while populations and country incomes are exogenously prescription often using SSPs scenarios (O'Neill et al. 2014). The LandSyMM approach offers the unique opportunity to represent trade-offs, responses and cross-scale and international interactions within a dynamic system (Rounsevell et al. 2014). Greenhouse gas emissions from land-use change and land management feedback to simulated climate. Currently in progress is a detailed representation of forestry.

Climate Change Adaptation Applications

From a consumption perspective, LandSyMM captures dietary requirements and preferences at the country level and how consumption changes in response to income levels as well as endogenous country level food commodity prices. In addition to the detailed representation of land management practices (e.g. technical efficiency, fertilizer use, transport, losses during transport, fertilizer cost) at a granular scale, we can assess the effects of autonomous decisions from the food system's actors and their interactions with mitigation policies. For instance, to study the interactions between a country who might have suffered from shocks (e.g. flood, drought, yield shocks, price shocks, pandemic and cyber-attacks) and the rest of the world under growing climate change pressure. The changing risk of some of these shocks, such as drought and flood risk, can be simulated by processes endogenous to LandSyMM (Fig. 16.2). In addition, it is possible to implement policy levers such as international trade tariff barriers and agricultural subsidies into the model to investigate their impacts on the food systems, dietary requirement, land use, climate but also their impacts on the potential benefices of certain climate adaptation measures. The increase in plant productivity under higher climate forcing intensity will change the production patterns at fine scale and creates new opportunities within the food system. The change in production may be related to the adoption of new crop types, the changes in management practices or the shift toward other cultivated areas (Alexander et al. 2018). In some cases, changing agricultural productivity could drive food substitutions in consumer choice leading to changes in a country's imports, exports and production without direct policy implications. Mitigation and adaptation to climate change do not always co-benefit. For instance, the widespread adoption of climate mitigation actions such as bioenergy and reforestation can impact the land and food systems, e.g. through the removal of existing agricultural land and increases in prices for agricultural commodities (Bahar et al. 2020).

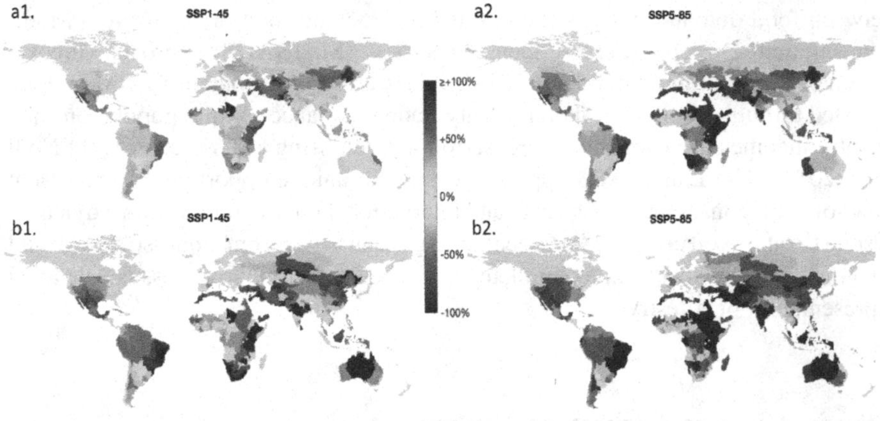

Fig. 16.2 LandSyMM output maps, aggregated to the basin scale, representing **a** the shift in the 5th percentile of annual surface runoff (drought risk), and **b** the shift in the 95th percentile of monthly surface runoff (flood risk) (**b**). The shift was calculated between the predicted values of 2071–2100 and 1971–2000 under (1) SSP 1 (a future in which challenges to both mitigation and adaptation are low) coupled with RCP 4.5 (climate given greenhouse gas emissions consistent with SSP1), and (2) under SSP 5 (a future in which challenges to mitigation are high and for adaptation are low) coupled with RCP 8.5 (climate given greenhouse gas emissions consistent with SSP5)

Conclusions and Recommendations

Agricultural adaptation measures are not necessarily positively synergistic with the environment. Autonomous adaptation in land manager choices is likely to be substantially driven by economic interests. As a result, land manager decisions are likely to minimize impact to market-based outputs, i.e. food and timber production. This means that the outcome for environmental externalities, including greenhouse gas emissions, fresh-water use and biodiversity loss may be detrimental (Fitton et al. 2019; Molotoks et al. 2020). Expansion of agriculture to new areas is expected to lead to carbon losses for soil and vegetation, while increases in food prices are likely to result in increased intensification of agricultural production, with negative environmental consequences. To date, there has been a lack of focus on the, potentially confounding, interactions between climate change mitigation and adaptation in the land system.

LandSyMM is an important new tool with the capacity to address this gap in understanding climate change adaptation-mitigation, and to inform policymakers on the trade-offs between different policy options as well as the impact on various aspects of the food system (e.g. production, international trade and diets). LandSyMM can be used to explore the potential of climate adaptation via the implementation of different scenarios that underpin climate change. These scenarios explore different futures such as different levels of economic growth, population demographics, international trade regimes and dietary preferences. The different scenarios enable the investigation of long-term impacts of policy measures on ecosystem services such as carbon storage, runoff, nitrogen losses, biogenic volatile organic compounds and biodiversity hotspots (Henry et al. 2019; Rabin et al. 2020). Despite the recent improvements, more research is needed to better reflect the reality of our complex world, e.g. integration of non-economic drivers for land manager decisions, and allowing bilateral trade to be represented. However, the current implementation of LandSyMM already provides a platform to better understand the interactions between land-based climate adaptation and mitigation that is currently lacking, and to identify suitable policies and actions.

Acknowledgements We acknowledge the support of the UK's Global Food Security Programme project Resilience of the UK food system to Global Shocks (RUGS, BB/N020707/1) and the Helmholtz Association.

References

Agnolucci P, Rapti C, Alexander P, Lipsis VD, Holland RA, Eigenbrod F, Ekins P (2020) Impacts of rising temperatures and farm management practices on global yields of 18 crops. Nat Food 1:562–571. https://doi.org/10.1038/s43016-020-00148-x

Alexander P, Rabin S, Anthoni P, Henry R, Pugh TAM, Rounsevell MDA, Arneth A (2018) Adaptation of global land use and management intensity to changes in climate and atmospheric carbon dioxide. Glob Chang Biol 24:2791–2809. https://doi.org/10.1111/gcb.14110

Bahar NHA, Lo M, Sanjaya M, Van VJ, Alexander P, Lckowitz A (2020) Meeting the food security challenge for nine billion people in 2050: What impact on forests? Glob Environ Chang 62. https://doi.org/10.1016/j.gloenvcha.2020.102056

Engström K, Olin S, Rounsevell MDAA, Brogaard S, van Vuuren DP, Alexander P, Murray-Rust D, Arneth A (2016) Assessing uncertainties in global cropland futures using a conditional probabilistic modelling framework. Earth Syst Dyn 7:893–915. https://doi.org/10.5194/esd-2016-7

Fitton N et al (2019) The vulnerabilities of agricultural land and food production to future water scarcity. Glob Environ Chang 58:101944. https://doi.org/10.1016/j.gloenvcha.2019.101944

Friedlingstein P et al (2019) Global carbon budget 2019. Earth Syst Sci Data 11:1783–1838. https://doi.org/10.5194/essd-11-1783-2019

Gouel C, Guimbard H (2018) Nutrition transition and the structure of global food demand. Am J Agric Econ 1–21. https://doi.org/10.1093/ajae/aay030

Henry RC, Alexander P, Rabin S, Anthoni P, Rounsevell MDA, Arneth A (2019) The role of global dietary transitions for safeguarding biodiversity. Glob Environ Chang 58:101956. https://doi.org/10.1016/j.gloenvcha.2019.101956

Humpenöder F, Popp A, Stevanovic M, Müller C, Bodirsky BL, Bonsch M, Dietrich JP, Lotze-Campen H, Weindl I, Biewald A, Rolinski S (2015) Land-use and carbon cycle responses to moderate climate change: Implications for land-based mitigation? Environ Sci Technol 49:6731–6739. https://doi.org/10.1021/es506201r

Huntingford C, Booth BBB, Sitch S, Gedney N, Lowe JA, Liddicoat SK, Mercado LM, Best MJ, Weedon GP, Fisher RA, Lomas MR, Good P, Zelazowski P, Everitt AC, Spessa AC, Jones CD (2010) IMOGEN: an intermediate complexity model to evaluate terrestrial impacts of a changing climate. Geosci Model Dev 3:679–687. https://doi.org/10.5194/gmd-3-679-2010

IPCC (2019) Climate Change and Land: an IPCC special report on climate change, desertification, land degradation, sustainable land management, food security, and greenhouse gas fluxes in terestial ecosystems

Molotoks A, Henry R, Stehfest E, Doelman J, Havlik P, Krisztin T, Alexander P, Dawson TP, Smith P (2020) Comparing the impact of future cropland expansion on global biodiversity and carbon storage across models and scenarios. Philos Trans R Soc B https://doi.org/10.1098/rstb.2019.0189

Müller B et al. (2020) Modelling food security: bridging the gap between the micro and the macro scale. Glob Environ Chang 63

O'Neill BC, Kriegler E, Riahi K, Ebi KL, Hallegatte S, Carter TR, Mathur R, van Vuuren DP (2014) A new scenario framework for climate change research: The concept of shared socioeconomic pathways. Clim Change 122:387–400. https://doi.org/10.1007/s10584-013-0905-2

Preckel PV, Cranfield JAL, Hertel TW (2010) A modified, implicit, directly additive demand system. Appl Econ 42:143–155. https://doi.org/10.1080/00036840701591361

Rabin SS, Alexander P, Henry R, Anthoni P, Pugh TAM, Rounsevell M, Arneth A (2020) Impacts of future agricultural change on ecosystem service indicators. Earth Syst Dyn 11:357–376. https://doi.org/10.5194/esd-2019-44

Robinson DT, Di Vittorio A, Alexander P, Arneth A, Barton CM, Brown DG, Kettner A, Lemmen C, Neill BC, Janssen M, Pugh TAM, Rabin SS, Rounsevell M, Syvitski JP, Ullah I, Verburg PH (2018) Modelling feedbacks between human and natural processes in the land system. Earth Syst Dyn 9:895–914. https://doi.org/10.5194/esd-2017-68

Robinson S, van Meijl H, Willenbockel D, Valin H, Fujimori S, Masui T, Sands R, Wise M, Calvin K, Havlik P, Mason d'Croz D, Tabeau A, Kavallari A, Schmitz C, Dietrich JP, von Lampe M (2014) Comparing supply-side specifications in models of global agriculture and the food system. Agric Econ 45:21–35. https://doi.org/10.1111/agec.12087

Rounsevell MDA, Arneth A, Alexander P, Brown DG, de Noblet-Ducoudré N, Ellis E, Finnigan J, Galvin K, Grigg N, Harman I, Lennox J, Magliocca N, Parker D, O'Neill BC, Verburg PH, Young O (2014) Towards decision-based global land use models for improved understanding of the Earth system. Earth Syst. Dyn. 5:117–137. https://doi.org/10.5194/esd-5-117-2014

Smith B, Wärlind D, Arneth A, Hickler T, Leadley P, Siltberg J, Zaehle S (2014) Implications of incorporating N cycling and N limitations on primary production in an individual-based dynamic vegetation model. Biogeosciences 11:2027–2054. https://doi.org/10.5194/bg-11-2027-2014

Chapter 17
Water Resource System Modelling for Climate Adaptation

Edoardo Borgomeo

Abstract Methods and models for water resource system simulation, risk analysis, and decision analysis provide powerful tools for dealing with the challenge of climate change in the water sector. These models enable learning about the complex behaviour of river basins, testing of alternative adaptation decisions, exploration of uncertainties, and navigation of trade-offs. This paper briefly describes recent advances in decision analysis and simulation modelling for climate adaptation in the water sector. These advances are now relatively mature and are increasingly being applied by practitioners.

Keywords Water resource system analysis · Water supply · Decision-making under uncertainty · Bottom-up vulnerability assessment · Optimization

Introduction

Methods for risk analysis and decision analysis provide powerful tools for informing climate change adaptation in the water sector. Simulation models are at the heart of these methods, and are widely used to improve understanding and assist decision-making in water resource management. The essence of water resource system simulation is predicting the hydrologic, socioeconomic, and environmental consequences of water management, especially in the face of climate change. These are the variables, such as future water availability, the economic value of water, and the reliability of environmental flows, that are important to governments, industries, and the public as they adapt to climate change (Brown et al. 2015). This paper briefly reviews the application of water resource system models for climate adaptation and sketches out some areas for further improvement. A water resources system is here defined as the whole made from connected hydrologic, infrastructure, ecologic, and human processes that involve water. It includes biogeophysical processes (e.g., elements of the hydrologic

E. Borgomeo (✉)
Environmental Change Institute, University of Oxford, 3 S Parks Rd, Oxford OX1 3QY, UK
e-mail: edoardo.borgomeo@ouce.ox.ac.uk

C. Kondrup et al. (eds.), *Climate Adaptation Modelling*, Springer Climate,
https://doi.org/10.1007/978-3-030-86211-4_17

cycle and ecosystem functioning), and human processes (e.g., construction, operation, and removal of infrastructure), and other human decisions and actions such as consuming, enjoying, being harmed by, or paying for water (Brown et al. 2015).

Decision Analysis for Water Management Under Climate Change

Faced with climate uncertainties, decision-makers may despair at the prospect of having to make long-term choices about adaptation and water infrastructure. These are indeed difficult decisions, which impact upon future generations and potentially lock-in patterns of development. The recognition of these challenges and of the impacts of climate change on water resources has given rise to a range of methods for decision-making under uncertainty (Maier et al. 2016). As we explain in the following section, all of these methods are underpinned by simulation models of hydrology and the operation of water resources systems (withdrawals, storage, allocation to users, and return flows). Before reviewing the characteristics of water simulators, we review a set of principles shared by these decision analysis methods:

First, they promote the notion of flexibility. Flexibility is broadly interpreted as the ability to switch or change a decision depending on what outcomes materialize. In practice, this means recognizing the extent to which modification of the operation of infrastructure systems can yield very different outcomes or building less infrastructure up front but enabling expansion in the future if needed.

Second, they promote the notion of robustness. A robust decision is a decision that performs acceptably well under a wide range of plausible future conditions. In the presence of uncertainty, it is desirable to seek water decisions that perform reasonably well across a range of possible future conditions, and so are robust to uncertainty. This emphasis upon robustness is in principle quite different to optimizing methods which focus on maximizing expected utility. Promoting a notion of robustness means identifying options that perform acceptably well (i.e., they satisfy a set of criteria) over the widest possible space of possible futures and not that are optimal over a narrow set of conditions.

Third, they emphasize the importance of exposing trade-offs in order to identify and mitigate possible undesirable impacts. The tools of multi-objective optimization are particularly powerful for exploring trade-offs between different attributes of water resource systems. These enable system states (e.g., in different possible future scenarios) to be presented in terms of their performance with respect to multiple objectives (Reed et al. 2013).

Simulation Models for Climate Adaptation in the Water Sector

Simulation modelling is at the core of the decision analysis methods described above. Simulation modelling is particularly powerful because of the capacity to test and explore shocks and scenarios that have never happened, by subjecting a simulator of a water resource system to those conditions 'in silico'. System stress testing through computer simulation is one of the most important tools for water resources planning in the face of climate change, especially when that is combined with scenario exercises for the multiple stakeholders responsible for a system. This approach, typically referred to as exploratory modelling, uses simulation models to ask 'what if' questions, unravelling the implications of different assumptions and hypotheses about future trends on water-related outcomes of concern (e.g., the frequency of water shortages) (Hall et al. 2019).

The purpose of water resource system simulators is to test the performance of the system under changing conditions of climate and demand. System performance within a given future state can be quantified in multiple ways. Since the earlier work of Hashimoto et al. (1982), who defined metrics of vulnerability, reliability, and resiliency, a number of other metrics have been applied to measure the performance of water resource systems. The recent emphasis on decision-making under uncertainty means that metrics that can be related to the principles of flexibility and robustness described above are increasingly being applied (e.g., maximin, optimism–pessimism, max regret) (Giuliani and Castelletti 2016). These metrics are typically quantified across large sets of plausible future climate scenarios (e.g., thousands of scenarios), generated either through downscaling of global climate model projections or through statistical models for direct simulation of hydroclimatic variables (e.g., synthetic hydrology). These metrics are often then traded-off against each other to identify acceptable decisions.

Simulating a water resource system involves coupling several different models (Hall et al. 2019):

Climatic boundary conditions: Information on climatic variables (rainfall, temperature) is a first key input to simulate water resource system behaviour under climate change. As described in Nazemi and Wheater (2014), this typically involves choosing one or more scenarios for future greenhouse gas concentrations to force one or more global climate models (GCMs) and then transferring the GCM projections of climate variables to the river basin of interest using one or more downscaling techniques. This process generates the climatic boundary conditions. Using climate models as the 'upstream' boundary of a water system simulator is attractive because (i) it enables simulations from climate models to be used (if necessary after appropriate downscaling) to test possible future climatic conditions and (ii) climate model outputs implicitly represent spatial and temporal dependencies between the several climatic variables that influence water resource systems—most notably precipitation and the variables that determine evapotranspiration. However, the reliability of GCM outputs for adaptation planning in the water sector has been questioned because of their inadequacy in terms of the scale of global and regional biases (Stainforth and Calel 2020). To overcome this challenge, climatic boundary conditions can also be generated through direct simulation of either weather (rainfall, temperature) (Steinschneider and Brown 2013) or hydrological (streamflow) variables (Borgomeo et al. 2015). Direct simulation of climatic boundary conditions is based on sampling and perturbation of the statistical distribution of historical observations.

Surface and groundwater hydrology: Hydrological models transform climatic inputs (notably rainfall) into quantities of water that may be withdrawn from rivers and/or groundwater at specified locations, taking into account topographical and land-use characteristics. Because the quantity of water at a given location is an aggregation of a complex series of spatial–temporal processes, these are dynamical models, though the representation of spatial complexity varies, from lumped catchment models to spatially explicit gridded models. These models are also increasingly capable of simulating water quality, which is known to influence water supplies for urban and rural users.

Water supply infrastructure (withdrawals, storage, pumping) and allocation rules: This is the core of a water resource system simulator. Water resource system models simulate the functioning of the water supply infrastructure, typically on daily or monthly timescales. System simulation models can represent rules for withdrawal of water from water bodies, operation of storage (e.g., dams), and allocation of water to different users. They take as input observations or projections of water demand and can also simulate the amount by which demand may be voluntarily or forcibly reduced during times of scarcity through water use restrictions.

Water use: A variety of methods exist for projecting water demands from households, agriculture, and other economic sectors. Traditionally, models focus on low time resolution data on consumption acquired through billing or limited measurement campaigns and employ deterministic forecast methods (House-Peters and Chang 2011). Recent advances in smart metering technology provide a promising avenue to advance residential water demand modelling and thus significantly improve the ability of water resource system simulators to model users' response to restrictions and other demand-side measures (Cominola et al. 2015). Advances in economic analysis also allow for an improved understanding and modelling of the impact of water use restrictions on multiple users (Freire-González et al. 2017). This understanding is crucial as it feeds directly into the cost–benefit assessment of policy options needed to reduce the risk of water shortages.

Adaptation Practices in the Water Sector: Simulating London's Water Security

The process illustrated above has been applied in London to adapt the city's water supply system to the expected impacts of climate change and population growth.

Simulators of London's water systems show that if no action is taken, London is indeed set to experience more frequent and severe water shortages in the future as early as 2030 (Borgomeo et al. 2018). This is mainly down to population growth, but climate change complicates things further as it will mean more frequent and intense droughts.

Through the use of simulators, water managers in London have identified aggressive demand management to reduce consumption and losses in the distribution system (called leakage) is a priority to be implemented immediately (Water 2019). However, they have also identified options to augment supplies in the long-term. These include recycling wastewater and transferring water from other parts of Southern England (Borgomeo et al. 2018; Water 2019). These options have been identified through the application of water resource system simulators designed to optimize system performance along four objectives: (1) least cost, (2) least environmental impact, (3) robustness, and (4) least emissions. London's approach to climate adaptation required water managers and regulators to move away from a decision model focused on identifying the least cost solution to close the water supply–demand gap. Instead, they expanded their decision objectives to incorporate other aspects such as sustainability and robustness which are key to adapt to climate change and which can be modelled through water system simulators.

The Expanding Boundaries of Water Resources System Modelling

Technological advances, institutional innovations, and behavioural change are some of the factors pushing the boundaries of water resource system modelling. Following Hall et al. (2020), we identify the following:

New data sources: The proliferation of sensors in water resource systems is providing an opportunity to fill persistent data gaps. For example, the introduction of smart water meters in homes is providing much more precise information on the characteristics of water usage.

Economics: There is a growing body of empirical research that seeks to quantify the interplay between water and the economy, in particular in economies that are highly dependent upon agriculture (usually, but not exclusively, poor countries) and are subject to large hydrological variability. Another strand of research examines the productivity of water and hence the wide economic effects of water shortages (Freire-González et al. 2017). This is a challenging research because water is so pervasive in the economy, so its effects are difficult to isolate.

Society: The study of the interplay between society and water has recently acquired the new title of socio-hydrology (Sivapalan et al. 2014). The emergence of socio-hydrology re-emphasizes a perennial need to better understand the complex human dimensions of water and incorporate these in the scientific analysis of water resource systems. These interactions operate at a very wide range of scales, from the choices made by individuals in households, through to the nature of water-related political conflicts in transboundary river basins.

Environment: Looking to the future, much more sophisticated understanding of the resilience of aquatic ecosystems is to be expected. This understanding can hopefully be used in a more dynamic way to inform water resources management decisions.

Conclusions and Recommendations

This paper reviews some of the current and future challenges and opportunities facing water resource system models for climate adaptation. Water resource system models and decision analysis methods are mature, and now see increasing application. However, there is still less uptake in practice than might be expected. This can be attributed to the approaches being conceptually quite challenging and not always easy to align with existing decision-making processes. On the other hand, intensifying calls for 'outcomes-based' management of water resources and for reporting of climate-related risks (e.g., climate-related disclosures in business) are now providing a powerful motivator for the wider adoption of these approaches.

References

Borgomeo E, Mortazavi-Naeini M, Hall JW, Guillod BP (2018) Risk, robustness and water resources planning under uncertainty. Earth's Future 6(3):468–487

Borgomeo E, Farmer CL, Hall JW (2015) Numerical rivers: a synthetic streamflow generator for water resources vulnerability assessments. Water Resour Res 51(7):5382–5405

Brown CM, Lund JR, Cai X, Reed PM, Zagona EA, Ostfeld A, Hall J, Characklis GW, Yu W, Brekke L (2015) The future of water resources systems analysis: toward a scientific framework for sustainable water management. Water Resour Res 51(8):6110–6124

Cominola A, Giuliani M, Piga D, Castelletti A, Rizzoli AE (2015) Benefits and challenges of using smart meters for advancing residential water demand modeling and management: a review. Environ Model Softw 72:198–214

Freire-González J, Decker C, Hall JW (2017) The economic impacts of droughts: a framework for analysis. Ecol Econ 132:196–204

Giuliani M, Castelletti A (2016) Is robustness really robust? How different definitions of robustness impact decision-making under climate change. Climatic Change 135(3–4):409–424

Hall JW, Borgomeo E, Bruce A, Di Mauro M, Mortazavi-Naeini M (2019) Resilience of water resource systems: lessons from England. Water Secur 8:100052

Hall JW, Borgomeo E, Mortazavi-Naeini M, Wheeler K (2019) Water resource system modelling and decision analysis. water science, policy, and management: a global challenge, pp 257–273

Hashimoto T, Stedinger JR, Loucks DP (1982) Reliability, resiliency, and vulnerability criteria for water resource system performance evaluation. Water Resour Res 18(1):14–20

House-Peters LA, Chang H (2011) Urban water demand modeling: review of concepts, methods, and organizing principles. Water Resour Res 47(5)

Maier HR, Guillaume JH, van Delden H, Riddell GA, Haasnoot M, Kwakkel JH (2016) An uncertain future, deep uncertainty, scenarios, robustness and adaptation: How do they fit together? Environ Model Softw 81:154–164

Nazemi A, Wheater HS (2014) Assessing the vulnerability of water supply to changing streamflow conditions. Eos, Trans Am Geophys Union 95(32):288–288

Reed PM, Hadka D, Herman JD, Kasprzyk JR, Kollat JB (2013) Evolutionary multiobjective optimization in water resources: the past, present, and future. Adv Water Resour 51:438–456

Sivapalan M, Konar M, Srinivasan V, Chhatre A, Wutich A, Scott CA, Wescoat JL, Rodríguez-Iturbe I (2014) Socio-hydrology: Use-inspired water sustainability science for the Anthropocene. Earth's Future 2(4):225–230

Stainforth DA, Calel R (2020) New priorities for climate science and climate economics in the 2020s. Nat Commun 11(1):3864

Steinschneider S, Brown C (2013) A semiparametric multivariate, multisite weather generator with low-frequency variability for use in climate risk assessments. Water Resour Res 49(11):7205–7220

Water T (2019) Draft water resources management plan. Thames Water plc., Reading UK

Chapter 18
A Top-Down Meets Bottom-Up Approach for Climate Change Adaptation in Water Resource Systems

Manuel Pulido-Velazquez, Patricia Marcos-Garcia, Corentin Girard, Carles Sanchis-Ibor, Francisco Martinez-Capel, Alberto García-Prats, Mar Ortega-Reig, Marta García-Mollá, and Jean Daniel Rinaudo

Abstract The adaptation to the multiple facets of climate/global change challenges the conventional means of water system planning. Numerous demand and supply management options are often available, from which a portfolio of adaptation measures needs to be selected in a context of high uncertainty about future conditions. A framework is developed to integrate inputs from the two main approaches commonly used to plan for adaptation. The proposed "top–down meets bottom–up" approach provides a systematic and practical method for supporting the selection of adaptation measures at river basin level by comprehensively integrating the goals of economic efficiency, social acceptability, environmental sustainability, and adaptation robustness. The top-down approach relies on the use of a chain of models to assess the impact of global change on water resources and its adaptive management over a range of climate projections. Future demand scenarios and locally prioritized adaptation measures are identified following a bottom-up approach through a participatory process with the relevant stakeholders and experts. Cost-effective combinations of adaptation measures are then selected using a hydro-economic model at basin scale. The resulting adaptation portfolios are climate checked to define a robust program of measures based on trade-offs between adaptation costs and reliability. Valuable insights are obtained on the use of uncertain climate information for selecting robust, reliable, and resilient water management portfolios. Finally, cost

---

M. Pulido-Velazquez (✉) · P. Marcos-Garcia · A. García-Prats
Research Institute of Water and Environmental Engineering (IIAMA), Universitat Politècnica de València UPV, Valencia, Spain
e-mail: mapuve@hma.upv.es

C. Girard
Fundació València Clima i Energia, Valencia City Council, Valencia, Spain

C. Sanchis-Ibor · M. Ortega-Reig · M. García-Mollá
Centro Valenciano de Estudios del Riego, UPV, Valencia, Spain

F. Martinez-Capel
Instituto de Investigación para la Gestión Integrada de Zonas Costeras IGIC UPV, Valencia, Spain

J. D. Rinaudo
BRGM, Montpellier, France

C. Kondrup et al. (eds.), *Climate Adaptation Modelling*, Springer Climate,
https://doi.org/10.1007/978-3-030-86211-4_18

allocation and equity implications are analyzed through the comparison of economically rational results (cooperative game theory) and the application of social justice principles.

Keywords Climate change adaptation · Water management · Robustness · Climate check · Top-down · Bottom-up

Introduction

Uncertainty and Adaptation in Water Resource Systems

The challenge of adaptation in water resource systems (WRS) includes coping with high/deep uncertainty about future resources ("end of stationarity") and demands. Water management problems, often classified as "wicked" management problems, involve dealing with multiple stakeholders with conflicting interests in a context of great complexity and shifting dynamics. There is a very broad range of potential adaptation options with different environmental and socioeconomic implications. Adaptation to climate/global change challenges the conventional means of water system planning, calling for a new paradigm in water management.

In any case, uncertainty cannot be an excuse for inaction. Flexible and dynamic adaption policies are to be set. Effective adaptation should combine both structural and non-structural measures, including regulatory and economic instruments. Selected adaptation is expected to be economically efficient, environmentally sustainable, socially acceptable, and robust. These are key requirements for the success of adaptation strategies. However, the integration of these factors in the decision-making process of the adaptation is a very complex issue still to be solved. This work presents a framework to include these attributes in the development of adaptation portfolios for river basins or WRS.

Top-Down Versus Bottom-Up Adaptation Strategies

Two main approaches are commonly implemented in the design of climate change adaptation plans. The "top-down" (TD) approach involves downscaling climate projections from General Circulation Models (GCM) under a range of emission scenarios to provide inputs for hydrologic and management models to estimate potential impacts and analyze adaptation measures. But this approach faces the problem of the "cascade of uncertainties", with uncertainty expanding at each step of the process when going from the global and regional projections to the study of the local impacts used to define the adaptation responses (Wilby and Dessai 2010).

Alternatively, in a "bottom-up" (BU) approach, vulnerability thresholds and local responses are empirically studied to define locally suitable adaptation strategies.

There are several interpretations of BU. Some authors refer to it when using local knowledge through participative approaches to foresight future scenarios and define locally relevant adaptation strategies (e.g., Bhave et al. 2014; Girard et al. 2015a), view adopted herein. Other authors consider BU as a scenario-free, robustness-based planning process; for example, in the "decision-scaling" approach (Brown et al. 2012; Poff et al. 2016; Ray et al. 2019). As for the later view, unlike the top-down method, the BU approach relies more on possibilities than on probabilities (Blöschl et al. 2013). However, this approach also depends on top-down information when assigning the probability to risky future climate conditions or selecting adaptation measures (e.g., Ray and Brown 2015).

Several authors have discussed the benefits of integrating TD and BU in the adaption process (e.g. Wilby and Dessai 2010; Ekström et al. 2013), although only a few studies have combined them in practice. We, herein, describe a framework for robust adaptation decision-making that departs from traditional methods, lying in the interface between the two aforementioned approaches. The purpose is the selection of portfolios of supply–demand measures for adaptation to climate change integrating the objectives of economic efficiency, environmental sustainability, acceptability, and robustness at basin scale.

Our views are shaped by recent experiences of developing adaptation strategies in two Mediterranean basins in France and Spain. In the Orb basin (1580 km^2), South-East France, climate change is expected to exacerbate the difficulty in meeting growing demands (high population growth and expectations of quick expansion of irrigated vineyards) while maintaining environmental in-stream flows. The management of the Jucar basin, Eastern Spain, larger (22,260 km^2), highly regulated, and with high share of water use for irrigation (around 80%), is already challenged by water scarcity and long recurrent multiannual (4–5 years) droughts.

Bottom-Up Approach

There are two main approaches for developing future land and water use scenarios for agriculture. One option is modeling land-use change (LUC) (e.g. Pulido-Velazquez et al. 2015). LUC modeling requires determining the drivers of change and spatial land use allocation applying machine learning techniques to historical observations. Using a combination of neural networks and cellular automata that learns from the past, we can translate regional projections from global scenarios into a map of future agricultural land use. The other option is the use of participatory approaches, involving the relevant actors through scenario-building workshops to develop plausible alternative futures (e.g., Rinaudo et al. 2013; Faysse et al. 2014).

Developing Future Demand Scenarios Through Scenario Building Workshops

Qualitative or quantitative approaches can be applied for the development of future scenario through a participatory approach. Qualitative scenarios can be useful for generating ideas and strategies and incorporating multiple viewpoints, bridging gaps among experts, decision-makers, and stakeholders Quantitative land-use scenarios, in contrast, describe plausible futures using numerical descriptions and spatial allocations of land uses associated with a potential pathway (Mallampalli et al. 2016). We adopt a mixed approach, using narrative texts (storylines) and translating them into quantitative scenarios. Next, the impacts triggered by the expected changes are assessed through model simulations. There is a broad range of methods for translating narrative scenarios into quantitative assessments of land use change (Mallampalli et al. 2016).

To identify future irrigation water demand in the Orb case study under climate change, we first defined future scenarios of land use changes through workshops. Agroclimatic simulation models were then used to determine the changes in irrigation needs (Girard 2015). Monthly average water demands were computed for nine climate projections. Future urban demand was also estimated using an econometric model, based on population, average household income, price, and climate.

As agriculture is by far the main water use in the Jucar basin, the characterization of future scenarios of this sector is crucial for water management. A first round of expert interviews were carried out to identify main drivers and trends in the agricultural sector in the basin. The interviews were helpful for adapting the main elements of the narratives of selected global Shared-Socioeconomic Pathways (SSPs) to the local context. The SSPs describe potential socioeconomic futures addressing different challenges in relation to both mitigation and adaptation policies (O'Neill et al. 2017). We conducted two focus workshops with representatives from the local agricultural sector in the two main agricultural areas to discuss two contracting SSPs global scenarios: SSP3 (regional conflicts, reversed globalization trends, with high challenges for both mitigation and adaptation) versus SSP5 (accelerated globalization, with low challenges for adaptation but high for mitigation). Global narratives were translated into local storytellings and depicted as fake future (2030) news in two local newspapers (Ortega-Reig et al. 2018). Local participation was key for developing an integrated vision of the evolution of agriculture and implications for water management in the context of the two SSPs and the climate change conditions corresponding to RCP 8.5. Changes in crop types, irrigated crop areas, and irrigation practices were discussed in accordance with the future socioeconomic and climate conditions presented to the participants. The associated changes in irrigation water requirements were estimated using crop simulation models considering climate change impact, which allowed to determine future water demand for the region.

Developing Portfolios of Water Management Adaptation Options at the Basin Scale

For the Orb river basin, after developing scenarios about the most likely evolution of urban and agricultural water use in the basin by 2030, possible adaptation measures were screened (Girard 2015). A first catalog of measures was elaborated by combining literature review and personal communications with consultation workshops involving local experts and stakeholders. Planned adaptation included optimization of reservoir operation, further development of groundwater, desalination, improved efficiency of large public agriculture irrigation schemes, leakage reduction in municipal water distribution networks, and implementation of tariffs as water conservation incentives (Girard et al. 2015a). Autonomous adaptation included water conservation actions at households, municipal services, and commercial activities under incentives. The stakeholder consultation process led to the identification of a list of priority measures (462 possible local measures of 13 types), while other measures were discarded (e.g., rainwater harvesting, wastewater reuse) based on technical, economic, legal, or acceptability criteria.

A participatory approach was also used in the Jucar basin for developing the portfolio of adaptation options for future scenarios. The suitability at basin scale of the adaptation measures previously proposed by the farmers was discussed at a third workshop that involved representatives of the main stakeholders in the basin (policymakers, users from agriculture, urban and hydropower sectors, environmentalist groups, etc.). After introducing each adaptation measure, participants discussed feasibility and potential implementation barriers, and graded each measure (both quantitatively and qualitatively) using an interactive participatory presentation platform through their mobiles. The qualitative assessment defined each measure as priority or supplementary, and identified potential-related issues (environmental impacts, social support, lack of training, political divisiveness, funding, effectiveness, and operational cost). Each measure was graded by the participants in a 0–10 scale (where 0 was meant for rejection) (Marcos-Garcia 2019). The measures consist of a new desalination plant, a wastewater reuse project, substitution of pumping by surface water in Mancha aquifer, and increase in irrigation efficiency by modernization (from flood to drip irrigation). Each measure was characterized in terms of water yield (effectiveness) and cost.

Top-Down Impact Assessment

The top–down approach starts by selecting a set of climate projections considering several emission scenarios and GCMs Models to account for uncertainty. These climate projections are then downscaled and bias-corrected to construct local climate change projections using dynamic or statistical downscaling techniques.

Local climate change projections are used as input to hydrological models to simulate the impact on the available resources. The local climate projections are also the input for the agro-climatic models.

For the Orb case, we used climate scenarios downscaled from nine GCMs. In order to capture the range of impacts introduced by climate change, results of all climate projections were considered equally likely. Large variations were observed in the results for the different climate models. A monthly lumped two-parameter rainfall-runoff model, forced by historical climatic data (precipitation and potential evapotranspiration) was calibrated and validated on each of the 11 sub-basins using the observed monthly discharge (Girard 2015).

For the Jucar case, combinations of GCMs-RCMs for the case study where selected by comparing observed versus simulated time series of mean annual precipitation and temperature for the control period (1971–2000). Hydrological changes were obtained from a Temez rainfall-runoff model modified to improve the simulation of stream–aquifer interaction. The resulting inflows in the climate change scenarios showed great variability across GCM/RCM model combinations, revealing high uncertainty in future water availability. Results also highlighted the spatial variability of climate change impacts in the basin. Temperature increase and precipitation decrease would be higher in the upper basin, where most reservoir storage capacity is located. Both meteorological and hydrological droughts are expected to grow in intensity, magnitude, and duration (Marcos-Garcia et al. 2017).

Integrating Top-Down and Bottom-Up Approaches

Monthly inflow time series for each climate projection at each subbasin obtained from the top-down approach and adaptation measures selected in the BU were integrated into a water management model used as decision support system (DSS) for the definition of adaptation strategies to climate change. The DSS consisted in a hydroeconomic model of the basin that, through optimization, selects the most cost-efficient combination of adaption measures for future scenarios. Hydro-economic models enable the definition of economically efficient adaptation by integrating hydrologic, engineering, environmental, and economic aspects of water resource systems within a coherent framework (Harou et al. 2009). They have been applied to assessing climate change impacts and the value of adaptation strategies for water systems (e.g. Escriva-Bou et al. 2017).

In the Orb basin, a river basin optimization model was used to select the combination of adaptation measures that minimizes the total annualized cost of adaptation while meeting the demand and minimum in-stream flow targets (Girard et al. 2015a, b). Constraints were defined to ensure certain reliability of deliveries to urban and agricultural demands and fulfillment of minimum environmental flow requirements. 11 subbasins, 64 urban and 19 agricultural demands were considered in the optimization model, which selected the optimal adaptation among 347 measures over 20 years of future monthly inflow. Optimal portfolios of lower cost measures were obtained for each future climate and land use scenario. The different portfolios of measures were characterized in terms of cost and reliability. In order to test the robustness of the optimal strategies, the performance of each of the nine portfolios was tested across the other climate projections, considering tradeoffs between adaptation cost and reliability of supply to agricultural demands. A multicriterion method was used to identify the most robust and least regretful solutions (Girard et al. 2015a).

In the Jucar basin, a water management hydroeconomic model integrating environmental restrictions, allocation rules (in accordance with Spanish and river basin regulations), and existing agreements was used to identify economically efficient adaptation strategies. For most climate scenarios, the selected measures allow to significantly reduce the average annual water deficit in the system.

Addressing Equity in Cost Allocation

Stakeholders will only agree to implement actions prescribed by a cost-effective plan if perceived as equitable. Cost-allocation scenarios were first designed by applying cooperative game theory based on the principle of economic rationality. The results were then contrasted with cost allocation scenarios representing alternative principles of social justice, investigated through semi-structured interviews with key local actors to obtain insights on the definition of a fair allocation of adaptation cost within the basin (Girard 2015). The comparison of the cost allocation scenarios led to contrasted insights to inform the decision-making process and potentially reap the efficiency gains from cooperation in the design of river basin adaptation portfolios (Girard et al. 2016).

Conclusions and Recommendations

The main contribution of this work is the development of a framework to identify adaptation options to climate change at the basin through the combination of a top-down (TD) approach to assess climate change impacts at the local scale with vulnerability assessment and definition of socioeconomic scenarios and adaptation options through participative methods (BU approach). The proposed "TD meets BU" approach provides a systematic and practical method for supporting the selection of adaptation measures at the basin by comprehensively integrating the goals of economic efficiency (through river basin optimization), social acceptability (through BU definition of scenarios and measures, and by addressing equity in cost allocation), environmental sustainability (through environmental constraints in water management), and robustness (testing robustness of adaptation portfolios across scenarios, and selecting robust/least-regret programs).

The "scenario foresight" approach has been shown to be useful for a BU exploration of local alternative futures. Experts and farmers have helped to analyze in a structured way the consequences of various global scenarios of climate and socio-economic change on future agriculture in a local context, and identify adaptation measures. Scenario workshops can usefully supplement modeling methods in the design and assessment of climate and global change scenarios and the selection adaptation strategies.

References

Bhave AG, Mishra A, Raghuwanshi NS (2014) A combined bottom-up and top-down approach for assessment of climate change adaptation options. J Hydrol 518:150–161

Blöschl G, Viglione A, Montanari A (2013) Emerging approaches to hydrological risk management in a changing world. In: Pielke RA (ed) Climate vulnerability. Academic, Cambridge, pp 3–10

Brown C, Ghile Y, Laverty M, Li K (2012) Decision scaling: linking bottom-up vulnerability analysis with climate projections in the water sector. Water Resour Res 48:W09537

Escriva-Bou A, Pulido-Velazquez M, Pulido-Velazquez D (2017) The economic value of adaptive strategies to global change for water management in Spain's Jucar basin. J Water Resourc Plan Manag 143(5):1–1

Ekström M, Kuruppu N, Wilby RL, Fowler HJ, Chiew FHS, Dessai S, Young WJ (2013) Examination of climate risk using a modified uncertainty matrix framework—applications in the water sector. Glob Environ Change 23:115–129

Faysse N, Rinaudo JD, Bento S, Richard-Ferroudji A, Errahj M, Varanda M, Imache A, Dionnet M, Rollin D, Garin P, Kuper M (2014). Participatory analysis for adaptation to climate change in Mediterranean agricultural systems: possible choices in process design. Reg Environ Chang 14(1):57–70

Girard C (2015) Integrating top-down and bottom-up approaches to design a cost-effective and equitable programme of measures for adaptation of a river basin to global change. PhD thesis, Universitat Politecnica de Valencia, Valencia, Spain. https://doi.org/10.4995/Thesis/10251/59461.

Girard C, Pulido-Velazquez M, Rinaudo JD, Pagé C, Caballero Y (2015a) Integrating top–down and bottom–up approaches to design global change adaptation at the river basin scale. Glob Environ Chang 34:132–146

Girard C, Rinaudo JD, Pulido-Velazquez M, Caballero Y (2015b) An interdisciplinary modelling framework for selecting adaptation measures at the river basin scale in a global change scenario. Environ Model Softw 69:42–54

Girard C, Rinaudo JD, Pulido-Velazquez M (2016) Sharing the cost of a river basin adaptation portfolios to climate change: insights from social justice and cooperative game theory. Water Resour Res 52(10):7945–7962

Harou J, Pulido-Velazquez M, Rosenberg DE, Medellín-Azuara J, Lund JR, Howitt RE (2009) Hydro-economic models: concepts, design, applications and future prospects. J Hydrol 375:627–643

Mallampalli VR, Mavrommati G, Thompson J, Duveneck M, Meyer S, Ligmann-Zielinska A, Druschke CG, Hychka K, Kenney MA, Kok K, Borsuk ME (2016) Methods for translating narrative scenarios into quantitative assessments of land use change. Environ Model Softw 82:7–20

Marcos-Garcia P, Lopez-Nicolas A, Pulido-Velazquez M (2017) Combined use of relative drought indices to analyze climate change impact on meteorological and hydrological droughts in a Mediterranean basin. J Hydrol 554:292–305

Marcos-Garcia P (2019) Sistema de ayuda a la decisión para la adaptación y gestión de sistemas de recursos hídricos en un contexto de alta incertidumbre. Aplicación a la cuenca del Júcar. PhD thesis, Universitat Politecnica de Valencia, Valencia, Spain

O'Neill BC, Kriegler E, Ebi KL, Kemp-Benedict E, Riahi K, Rothman DS, Van Ruijven BJ, van Vuuren DP, Birkmann J (2017) The roads ahead: narratives for shared socioeconomic pathways describing world futures in the 21st century. Glob Environ Chang 42:169–180

Ortega-Reig M, García-Mollá M, Sanchis-Ibor C, Pulido-Velázquez M, Girard C, Marcos P, Ruiz-Rodríguez M, García-Prats A. (2018). Adaptation of agriculture to global change scenarios. Application of participatory methods in the Júcar River basin (Spain) (in Spanish). Economía Agraria y Recursos Naturales 18(2):29–51.

Poff NL, Brown CM, Grantham TE, Matthews JH, Palmer MA, Spence CM, Wilby RL, Haasnoot M, Mendoza GF, Dominique KC, Baeza A (2016) Sustainable water management under future uncertainty with eco-engineering decision scaling. Nat Clim Chang 6(1):25–34

Pulido-Velazquez M, Peña-Haro S, Garcia-Prats A, Mocholi-Almudever AF, Henriquez-Dole L, Macian-Sorribes H, Lopez-Nicolas A (2015) Integrated assessment of the impact of climate and land use changes on groundwater quantity and quality in Mancha Oriental (Spain). Hydrol Earth Syst Sci 19:1677–2169

Rinaudo JD, Maton L, Terrason I, Chazot S, Richard-Ferroudji A, Caballero Y (2013) Combining scenario workshops with modeling to assess future irrigation water demands. Agric Water Manag 130:103–112

Ray PA, Brown CM (2015) Confronting climate uncertainty in water resources planning and project design: the decision tree framework. The World Bank, Washington D.C.

Ray PA, Taner MÜ, Schlef KE, Wi S, Khan HF, Freeman SSG, Brown CM (2019) Growth of the decision tree: advances in bottom-up climate change risk management. J Am Water Resour Assoc 55(4):920–937

Wilby RL, Dessai S (2010) Robust adaptation to climate change. Weather 65:180–185

Chapter 19
Advances in Climate Adaptation Modeling of Infrastructure Networks

Raghav Pant

Abstract As the adverse effects of climate change are increasingly becoming unavoidable, calls for improving climate adaptation assessments have gathered interest at the global scale. Infrastructure policymakers and practitioners are now interested in understanding climate vulnerabilities and risks that capture the systemic nature of failure propagation seen across interconnected networks. This would help inform adaptation planning objectives meant to improve systemic resilience. This paper presents recent technical methodological and tool-based advances made in climate vulnerability, risk, and adaptation modeling of large-scale infrastructure networks. These methodologies adopt a bottom-up approach that focuses on creating data-rich representations of infrastructure network attributes, resource flows, and socio-economic indicators that are all used for quantifying direct and indirect risks to network assets exposed to extreme climate hazards at multiple scales. Insights from different case studies are presented to show how such methodologies have been used in practice for informing different policy needs. The paper concludes by identifying the existing gaps and future opportunities for such bottom-up infrastructure network vulnerability, risk, and adaptation assessment methodologies.

Keywords Infrastructure networks · Climate change · Vulnerability · Risks · Adaptation

Introduction

There is increasing evidence that climate change will increase variability of weather patterns, magnifying the severity of short-term shock events such as flooding, storms, heatwaves, while also extending long-term shock events such as droughts (Wang et al. 2017). While mitigation efforts to limit the increase in global average temperature to 1.5 °C above pre-industrial levels remain a primary focus of policymakers (Christoff 2016), there is also increasing awareness that climate change adaptation action is

R. Pant (✉)
Environmental Change Institute, University of Oxford, Oxford, UK
e-mail: raghav.pant@ouce.ox.ac.uk

C. Kondrup et al. (eds.), *Climate Adaptation Modelling*, Springer Climate,
https://doi.org/10.1007/978-3-030-86211-4_19

needed with some urgency (Global Commission on Adaptation 2019). As stated by the European Commission (2014) "Adaptation means anticipating the adverse effects of climate change and taking appropriate action to prevent or minimize the damage they can cause or taking advantage of opportunities that may arise". Among others, climate adaptation is becoming an important focus of national and global infrastructure system planners, investors, and decision-makers who are faced with new challenges of embedding adaptation planning into objectives of managing sustainable development, economic prosperity, and demands of growing population. In particular, climate adaptation of economic infrastructures, which include large-scale spatially distributed networks of energy, transport, water, waste, telecommunications, is now a key topic of interest because such networks are recognized as lifelines of modern societies (Hallegatte et al. 2019).

The focus of this paper is on building data-driven decision-making models and tools for evaluating the costs and benefits of climate adaptation of infrastructure networks at different spatial scales. There has been an increasing demand for such tools from policymakers and practitioners interested in improving their decision-making for monitoring and evaluating adaptation options (European Commission 2014). When it comes to infrastructures, some questions relevant to informing climate adaptation include: (1) What are the key network locations and assets exposed to current and future climate-change driven hazards? (2) How do asset vulnerabilities and risks cascade across infrastructure networks? (3) What are the indirect consequences of network failures in terms socio-economic impacts felt beyond the initiating infrastructures? (4) What are some key climate adaptation investments and strategies for reducing network risks? (5) Where and what are the key infrastructure network locations prioritized for climate adaptation measures to reduce systemic network risks? This paper discusses how the above questions are being answered with generalized methodologies supported by data-driven case studies in different countries and at the global scale. These methodologies and case studies are all from the experiences and examples of work done in the Infrastructure Transition Research Consortium (ITRC 2020), which aims to build data-driven models for the identification of spatial network vulnerabilities and risks to support decision-making. Through these studies, the paper highlights the specific issues, recent advances, and further opportunities in creating technical knowledge to improve the information of climate adaptation modeling of infrastructure networks.

Methodologies for Evaluating Network Vulnerabilities, Risks, and Adaptation

A generalized methodology or framework for climate vulnerability, risk, and adaptation assessments would be difficult to conceive, as it is nearly impossible to account for every context-specific issue. Nonetheless, some broad principles of such a framework would include, among others: (1) identifying current and future climate vulnerabilities and risks under different climate hazard scenarios; (2) integrating climate risks within the decision-making process; and (3) identifying options and prioritizing responses based on the benefits of implementing such options. Most approaches that incorporate these steps apply a top-down modeling philosophy, where the identification of vulnerabilities and risks is mainly done by quantifying the direct physical impacts induced by exposures to external hazard shocks and any uncertainty in estimates is purely a function of the variability of climate model outputs (Conway et al. 2019). An example of such an approach is a global-scale assessment of lengths (in kilometers) and losses in asset damage costs (in US$) estimated for roads and railway assets exposed to multi-hazard risks (Koks et al. 2019).

More relevant for infrastructure network vulnerability, risk, and adaptation assessments are methodologies that also take a bottom-up modeling approach, which incorporates information at finer geographic scales with underlying physical, operational, social, and economic aspects associated with systems (Conway et al. 2019). Infrastructures operate as a "system-of-systems" of interdependent networks that are increasingly reliant on each other for services under normal working conditions, but which also create failure cascades from an originating asset toward the rest of the system-of-systems (Hall et al. 2019). Most climate vulnerability, risk, and adaptation analysis studies fail to capture the sensitivities of cascading failure mechanisms across infrastructure networks along with the sensitivities associated with weather and climate extremes. The methodological steps toward a hybrid top-down and bottom-up climate vulnerability, risk, and adaptation assessment of infrastructure networks involve combining (Hall et al. 2019; Pant et al. 2020): (1) Climate hazard information that includes spatially correlated probabilities, magnitudes, and extents of hazard events under current and future climate scenarios; (2) spatial representations of network point and line assets to evaluate their exposures to various climate hazards; (3) direct vulnerability measures of the exposed network assets that quantifies the sensitivity of the assets to be damaged by varying severities of climate hazards; (4) network connectivity effects that capture how failures cascade from the directly damaged assets toward other assets either physically or through the flow of resources (goods, information, etc.); (5) indirect vulnerability measures that quantify the socio-economic effects resulting from disruptions of infrastructure network services in terms of numbers and monetary values attached to household and business customers; (6) further indirect vulnerability measures of impacts to the regional economic flows that quantify the effects of production and labour disruptions on the outputs of macroeconomic sectors; (7) measurements of risks as the product of the probabilities, exposures and direct and indirect vulnerabilities summed over all possible hazard and network failure and disruption scenarios; (8) quantifiable options for building resilience (to climate or any shock event) of individual assets and the networks that include, but are not limited to, upgrading existing design standards of assets to withstand more extreme shocks, incorporating backup options to substitute for disruptions of services provided from one network to another (e.g. electricity backup generators at railway stations), increasing network redundancy and rerouting options to maintain resource flows, speeding up the recovery of damaged assets to bring back the networks to normal levels of service.

The effectiveness of different resilience options (mentioned in Step 8 above) is evaluated in terms of their costs and the benefits of avoiding direct and indirect risks of failures. Climate adaptation assessment involves the consideration of changing costs and benefits (avoided risks) over the life cycle of asset and network management. The costs over the lifetime include (Oh et al. 2019; Pant et al. 2019): (1) Initial investment costs of adaptation which are the one-time costs of a resilience option when it is implemented and (2) costs of routine maintenance (assumed to apply every year) and periodic maintenance (assumed to apply every few years) over the life cycle of the asset. To inform decision-makers on how to prioritize investment decisions for adaptation planning, a cost–benefit analysis is done to estimate network asset and location-specific benefit–cost ratios (BCR) that help identify for which options the avoided asset risks would be worth implementing ($BCR \geq 1$) and which options would be more expensive than the avoided risks ($BCR < 1$).

The purpose of the methodology outlined above is to create an effective high-level screening whereby the efficacy of different types of adaptation options can be compared and locations and assets with high (or low) adaptation benefits can be narrowed down for further investigation. At a country scale, where it is not possible to invest in every asset to make it climate resilient, such information is useful for narrowing down the locations of risks, which should be followed by detailed site-specific investigations of, among others, local conditions of hazards and assets risks.

A few case study examples are presented next to highlight how the described methodology has been implemented in practice to inform decision-makers about quantifiable vulnerabilities and risks to infrastructure networks at the national scale.

Case Studies of Infrastructure Vulnerability, Risk, and Adaptation Assessments

To understand the cascading nature of network failures, a recent study of interdependent energy, water, waste, telecommunications, and transportation network failures in New Zealand, to inform Civil Defence Emergency Management, showed that from all simulated failure events, nearly half (46%) of the total disruptions could be attributed to network propagation effects instead of disruptions attributed to the directly failed assets (Zorn et al. 2020). Also, recent analysis on infrastructure resilience in Great Britain, done for the National Infrastructure Commission, looked at interdependent electricity and telecom networks with dependent water, rail, and road networks (Pant et al. 2020). The study showed that if failures were initiated in the electricity network, then about 40% of failure events led to further disruptions to telecoms and at least one of rail and water networks, which set up another sequence of failures where 20% of failure events led to further electricity failures, and 5.7% to another order of telecoms failures. But if failures were initiated in the telecom network, then only about 7.8%

of failure events led to electricity and at least one of the rail and water disruptions, with 1.8% events leading to further sequence of telecom failures. The study also explored different resilience combinations of installing backup electricity supply for limited durations and increasing network redundancies, which on average reduced the worst-case socio-economic disruptions of network failures by 89%–94%. While they do not include any climate hazards and risks, these studies have proved effective in informing stakeholders how interdependencies influence failure cascades, especially creating feedbacks that lead to further disruptions in the networks where the failures originate. The analyses highlighted the regularity with which failures in networks like electricity could lead to several orders of cascading failures, which was not demonstrated previously in these countries.

Studies done with the World Bank in Tanzania (Pant et al. 2018), Vietnam (Oh et al. 2019), and Argentina (Pant et al. 2019) on multi-modal transport networks have estimated risks due to failures of key network links exposed to one or more climate hazards (e.g. floods, cyclones, landslides). These studies have helped inform transport investors and policymakers in these countries about the locations of their critical transport assets and routes, by estimating and comparing the ranges of magnitudes of freight tonnage disruptions and macroeconomic losses incurred from asset failures. By incorporating the sensitivities of climate models and scenarios with transport and economic growth forecasts, the studies have informed stakeholders about the changing risks in the future. For example, the Tanzania analysis estimated that the worst-case transport asset failure would result in US$ 1.4 million/day economic losses in 2018 due to supply chain disruptions and by 2030, the same failure scenario would create as high as US$ 2.5 million/day economic losses under a 6%–7% GDP growth forecast (Pant et al. 2018).

The studies in Vietnam and Argentina also incorporated adaptation options of upgrading roads and bridges to higher climate resilience designs and assessed the BCRs of such adaptation options. Both studies highlighted that the transport networks in these countries were increasingly exposed to more severe and frequent extreme hazards, and their respective road networks would require significant investments to upgrade to higher climate-resilient design standards. In Vietnam estimates, it was suggested that adaptation investment costs of upgrading the 20 worst-impacted national roads would be high, but the cumulative benefits over 35 years of such investments were substantial, where for every 1 US$ invested in enhancing climate resilience, the benefits of avoiding risks would be equivalent to safeguarding US$ 7–23 of economic value associated with freight supply chains (Oh et al. 2019). Similarly, in Argentina, for building resilience to flood risks, the costs of investments could be high, but the benefits of avoiding damages and disruption losses would outweigh the investment costs (Pant et al. 2019). Furthermore, the case for investing in climate resilience became stronger as the durations of disruptive impacts increased, which in turn meant that investing in reducing the duration of disruption should be a priority of adaptation planners in Argentina (Pant et al. 2019). These studies are proving useful in providing scientific evidence to the transport ministries to develop a national strategy for climate-resilient transport and plans, as part of the transport sector's contribution to the Nationally Determined Contributions, to meet the Paris Climate Agreement targets (Oh et al. 2019). The results of the analyses are also being provided as web-based tools for data inquiry and detailed network-scale climate risks and adaptation outcomes, which are integrated into the data analytic systems being used by the Ministry of Transport in Argentina (Pant et al. 2019).

Tools Developed Through Case Studies

In the process of implementing the case studies discussed above, several open-source Python programming-based tools have been created. We discuss three Python resources from the transport risk and adaptation analysis studies done in Vietnam and Argentina.

For the Vietnam study, the GitHub repository (https://github.com/oi-analytics/vietnam-transport) was created to host the Python codebase for Vietnam-specific analysis. A similar Python codebase was also developed for the transport analysis in Argentina, available at (https://github.com/oi-analytics/argentina-transport). Both these codebase repositories allow the user with information to install a set of functions for creating spatial analysis functions of processing hazard datasets, create networks, perform transport network flow and failure analysis, and do the adaptation analysis. For the Argentina analysis, a web-based risk visualization platform, available at https://github.com/oi-analytics/oi-risk-vis, was also developed as a results inquiry tool to help stakeholders identify and zoom in on the locations of the most vulnerable transport network assets in the country. The tool also helps identify locations of the roads and bridges where adaptation investments should be prioritized based on their BCRs, as shown in Fig. 19.1.

Conclusions and Recommendations

The hybrid top-down and bottom-up methodologies, for quantifying climate vulnerabilities, risks, and adaptation for infrastructure networks, discussed in this paper, are being increasingly required in making policy decisions. While policymakers and investors make infrastructure planning decisions at the asset scale, they lack knowledge on spatial network vulnerabilities and risks, resulting in investments being often not prioritized with the aim of building systemic resilience. But this gap is being filled as such network analyses are becoming more prevalent and achievable because more data on spatial hazards, spatial assets and network topologies, customer and economic usage are becoming available. Improving these methodologies and their uptake worldwide has been highlighted as one of the greatest adaptation opportunities (Hall et al. 2019).

The key gaps and opportunities lie in enhancing data collection to reduce the uncertainties in model estimates and improve confidence in the outputs of such analyses.

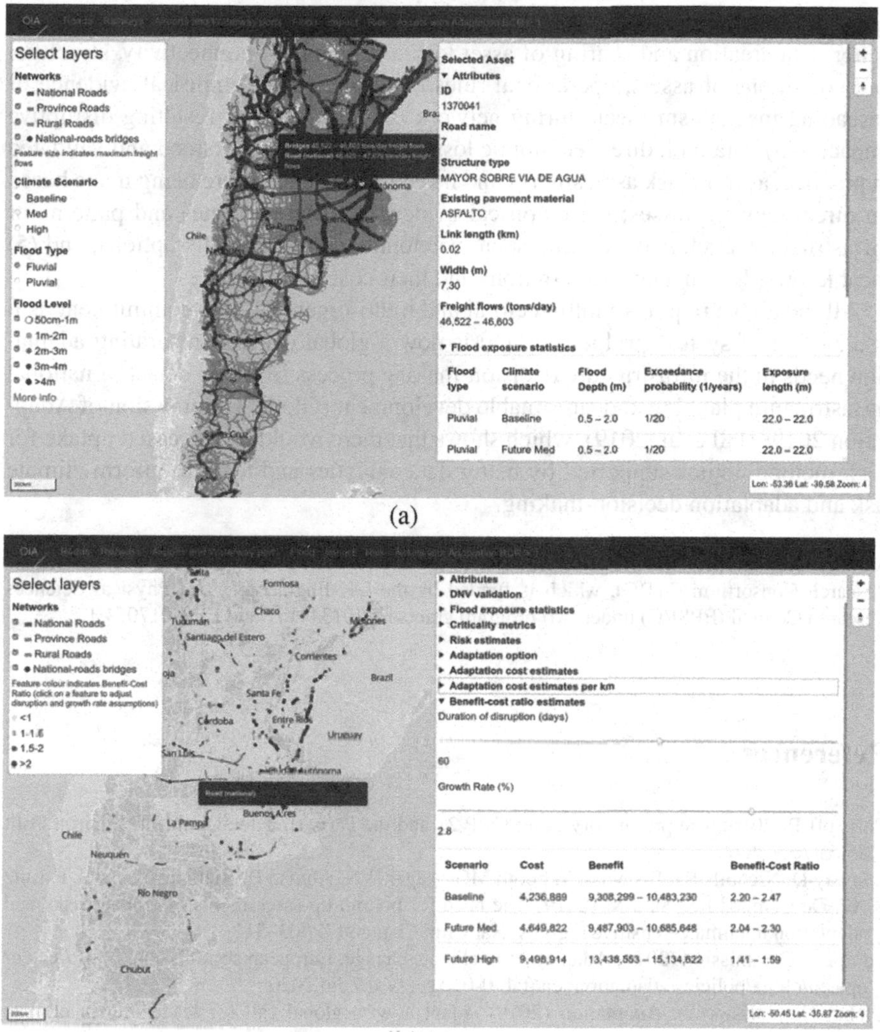

Fig. 19.1 Risk Visualization tool outputs at asset level for Argentina transport analysis study showing: **a** Characteristics and level of flood exposures of a road and **b** Road asset highlighted and identified by the BCRs of investing in climate adaptation

A lot of focus is given in improving hazard modeling, as it has been widely acknowledged that most climate hazard models are based on Global Climate Model scenarios at very coarse spatial scales, which lack regional climate information and show a lot of variability in their estimates, and fail to satisfactory represent some effects such as sea-level rise (Conway et al. 2019). Relatively less focus has been given on data collection on the impact assessment elements of the analyses, where a lot of opportunities for improvement lie. Hence, based on the experiences from different case studies, a key recommendation of this paper is to focus on, among others: (1) better data creation and sharing of asset locations, network connectivity, structural and conditions of assets, operational rules of networks; (2) empirical evidence of cascading mechanisms seen during network failures and their resulting disruptive impacts; (3) data on indirect economic losses that are less understood and validated in practice, as most risk assessments and investment decisions are being made based on direct damage losses; (4) empirical evidence on the timelines and patterns of infrastructure asset, network, and socio-economic recovery for disruptions; and (5) local knowledge on adaptation options and their costs.

All the above requires multi-sectoral and multi-organizational commitment as it is a system-of-system problem. There is now a global call for integrating adaptation needs in the infrastructure decision-making process from the outset of national infrastructure planning and sustainable development (Global Commission of Adaptation 2019; Hall et al. 2019), which shows that there would be increased uptake for such methodologies supported by better data analytics and tools to inform climate risk and adaptation decision-making.

Acknowledgements The author acknowledges the contribution to the Infrastructure Transitions Research Consortium (ITRC), which is funded by the UK Engineering and Physical Sciences Research Council (EPSRC) under two program grants EP/I01344X/1 and EP/N017064/1.

References

Christoff P (2016) The promissory note: COP 21 and the Paris climate agreement. Environ Polit 25(5):765–787

Conway D, Nicholls RJ, Brown S, Tebboth MG, Adger WN, Ahmad B, Biemans H, Crick F, Lutz AF, De Campos RS, Said M (2019) The need for bottom-up assessments of climate risks and adaptation in climate-sensitive regions. Nat Clim Chang 9(7):503–511

European Commission (2014) Adaptation to climate change, European commission. https://ec.europa.eu/clima/policies/adaptation_en#tab-0-0. Accessed 7 Jul 2020

Global Commission on Adaptation (2019) Adapt now: a global call for leadership in climate resilience. Global Centre on Adaptation, Netherlands

Hall JW, Aerts JCJH, Ayyub BM, Hallegatte S, Harvey M, Hu X, Koks EE, Lee C, Liao X, Mullan M, Pant R, Paszkowski A, Rozenberg J, Sheng F, Stenek V, Thacker S, Väänänen E, Vallejo L, Veldkamp TIE, van Vliet M, Wada Y, Ward P, Watkins G, Zorn C (2019) Adaptation of infrastructure systems: background paper for the global commission on adaptation. Environmental Change Institute, University of Oxford, Oxford

Hallegatte S, Rentschler J, Rozenberg J (2019) Lifelines: the resilient infrastructure opportunity. Sustainable infrastructure series. World Bank, Washington, DC. https://doi.org/10.1596/978-1-4648-1430-3. License: Creative Commons Attribution CC BY 3.0 IGO

ITRC (2020) UK infrastructure transitions consortium. https://www.itrc.org.uk/highlights/risk-resilience/. Accessed 11 Jul 2020

Koks EE, Rozenberg J, Zorn C, Tariverdi M, Vousdoukas M, Fraser SA, Hall JW, Hallegatte S (2019) A global multi-hazard risk analysis of road and railway infrastructure assets. Nat Commun 10(1):1–11

Oh JE, Espinet Alegre X, Pant R, Koks EE, Russell T, Schoenmakers R, Hall JW (2019) Addressing climate change in transport: volume 2: pathway to resilient transport. https://doi.org/10.1596/32412.

Pant R, Koks EE, Russell T, Hall JW (2018) Transport risks analysis for the United Republic of Tanzania – systemic vulnerability assessment of multi-modal transport networks. Final report draft, Oxford Infrastructure Analytics Ltd., Oxford, UK

Pant R, Koks EE, Paltan H, Russell T, Hall JW (2019) Argentina – transport risk analysis. Final report, Oxford Infrastructure Analytics Ltd., Oxford, UK

Pant R, Russell T, Zorn C, Oughton E, Hall JW (2020) Resilience study research for NIC – systems analysis of interdependent network vulnerabilities. Environmental Change Institute, Oxford University, UK. https://www.nic.org.uk/wp-content/uploads/Infrastructure-network-analysis.pdf

Wang G, Wang D, Trenberth KE, Erfanian A, Yu M, Bosilovich MG, Parr DT (2017) The peak structure and future changes of the relationships between extreme precipitation and temperature. Nat Clim Chang 7(4):268–274

Zorn C, Pant R, Thacker S, Shamseldin AY (2020) Evaluating the magnitude and spatial extent of disruptions across interdependent national infrastructure networks. ASCE-ASME J Risk Uncertain Eng Syst Part B: Mech Eng 6(2)

Chapter 20
Navigating Deep Uncertainty in Complex Human–Water Systems

C. D. Pérez-Blanco

Abstract Complex human–water systems are deeply uncertain. Policymakers are not aware of all possible futures (deep uncertainty type 2), while the probability of those futures that can be identified ex-ante is typically unknown (deep uncertainty type 1). In this context, standard decision-making based on a complete probabilistic description of future conditions and optimization of expected performance is no longer appropriate; instead, priority should be given to robustness, through the identification of policies that are (i) insensitive to foreseeable changes in future conditions (classical robustness that addresses deep uncertainty type 1) and (ii) adaptive to unforeseen contingencies (adaptive robustness that addresses deep uncertainty type 2). This research surveys recent advances in (socio-)hydrology and (institutional) economics toward robust decision-making. Despite significant progress, integration among disciplines remains weak and allows only for a fractioned understanding and partial representation of uncertainty. To bridge this gap, I will argue that science needs to further underpin the development and integration of two pieces of ex-ante information: (1) a modeling hierarchy of human–water systems to assess policy performance under alternative scenarios and model settings, so as to navigate deep uncertainty type 1 and (2) a longitudinal accounting and analysis of public transaction costs to navigate deep uncertainty type 2.

Keywords Deep uncertainty · Socio-hydrology · Robustness

Introduction

Climate change, population growth and changing distributions of wealth will lead water demand to outstrip supply by 40% in 2030, causing GDP growth to decline by as much as 6% in water-scarce areas (i.e. continued negative growth) (World Bank 2016). At the other extreme, floods represent the most economically damaging risk, costing circa $100 billion annually, and their impact is expected to rise to

C. D. Pérez-Blanco (✉)
Universidad de Salamanca and Centro Euro-Mediterraneo sui Cambiamenti Climatici, Salamanca, Spain
e-mail: Dionisio.perez@usal.es

C. Kondrup et al. (eds.), *Climate Adaptation Modelling*, Springer Climate,
https://doi.org/10.1007/978-3-030-86211-4_20

$521 billion/year in 2030 (World Resources Institute 2019). The combined effects of growing water scarcity and flood risk increasingly constrain decision-makers to adopt new approaches and policies to the management of the human-modified water cycle. *Critically*, the dynamics of complex human–water systems of relevance for water policy design and implementation are characterized by positive feedbacks, non-mechanistic dynamics and multiple equilibria leading to Knightian or **deep uncertainty**, where it is not possible to identify all possible futures (deep uncertainty type 2) or assign a probability to each identified possible future (deep uncertainty type 1). Under deep uncertainty, standard decision-making based on a complete probabilistic description of future conditions and optimization of expected performance is no longer appropriate; instead, **priority should be given to robustness**, through: (i) the avoidance of policies leading to unfavorable contingencies that can be identified beforehand (classical robustness, which addresses deep uncertainty type 1) (Marchau et al. 2019) and (ii) the avoidance of path-dependent trajectories, so to enable future adaptation to unpredictable, surprising, and potentially catastrophic ("black swan") events that are explainable only after they happen (adaptive robustness, which addresses deep uncertainty type 2) (Garrick 2015).

This research surveys recent advances in (socio-)hydrology and (institutional) economics that contribute toward uncertainty sampling and robust decision-making. I will argue that despite significant progress, integration among disciplines remains weak and allows only for fractioned understanding and partial representation of uncertainty. To bridge this gap, along these pages, I develop a research agenda toward an interdisciplinary, replicable, and scalable research framework integrating data and methods from (socio-)hydrology and economics to *quantify* the broad socioeconomic and environmental implications of adaptation policies in complex human–water systems, and the uncertainty involved in the process, so to allow stakeholders to explicitly trade-off incremental changes in robustness with expected policy performance (e.g. cost-effectiveness). To this end, I argue that science needs to further underpin the development and integration of two pieces of ex-ante information: (1) a modeling hierarchy of human–water systems to assess policy performance under alternative scenarios and models/model settings, so as to navigate deep uncertainty type 1 and (2) a longitudinal accounting and analysis of public transaction costs from before the project or policy commences. Public transaction costs are the institutional and organizational investments required to arrange, monitor, and enforce a policy and are instrumental to *measure* institutions' adaptive ability, avoid path dependent and potentially irreversible trajectories and strengthen adaptive robustness that addresses deep uncertainty type 2. Note that the first piece of information is only partially addressed in the scientific literature, which appears biased toward consolidative modeling and standard decision-making (Marchau et al. 2019); while empirical longitudinal assessments on public transaction costs are "virtually non-existent" (Loch and Gregg 2018).

Building a framework that addresses these gaps is challenging, but now feasible due to: (1) recent growth in availability of data from hydrology and socioeconomic domains (e.g. micro- and macro-economic); (2) recent advances in computational and statistical techniques for processing and harmonizing big data; (3) the growing number of water policy reforms, which can serve as 'living laboratories' for the collection, measurement, and analysis of public transaction costs (Garrick 2015); and (4) the consolidation of analytical and modeling methods proposed by emerging water resource research literature to study impacts and adaptation, including alongside stakeholders (Marchau et al. 2019).

Navigating Deep Uncertainty Type 1: Modular Hierarchies for Multi-system Ensembles

Three fundamental sources of deep uncertainty type 1 can be distinguished: (1) uncertainty arising from scenario assumptions and design (Marchau et al. 2019); (2) "parameter and structural uncertainties" within models (Tebaldi and Knutti 2007); and (3) uncertainty arising from the missing or "overly simplistic" representation of the interconnected dynamics of complex adaptive human–water systems (Pande and Sivapalan 2017).

The first two sources of uncertainty have been addressed with relative success. The Society for Decision Making under Deep Uncertainty has developed tools to address uncertainty arising from scenario assumptions and design through an exploratory modeling approach. Exploratory modeling and analysis works as a prosthesis for the intellect, using computational experiments representing the consequences of alternative sets of feasible assumptions to discover the implications of a priori knowledge—including domains of previously unforeseen contingencies. This information can then be used to illustrate relevant tradeoffs and revise scenarios and policy adoption in successive iterations leveraging on stakeholder and expert feedback until a robust policy is agreed upon (Marchau et al. 2019).

An ensemble of models can be used to sample uncertainty arising from parameter and structural uncertainties. Economic and hydrologic sciences have been successful at developing scientifically sound conceptual models capable of representing the essence of critical systems within the human–water conundrum. These include microeconomic models to represent the behavior of individuals or firms, macroeconomic models to study interrelations among sectors and regions of the economy and their impact on aggregated indicators and hydrologic models to study the movement, distribution, and quality of water at different scales, among other modeling families. There is consensus in the literature that the combination of scientifically sound prediction methods in perturbed physics and multi-model ensemble experiments (i.e. grouping multiple models and exploring alternative values for critical parameters) can be used to sample parameter and structural uncertainties through the ensemble spread. This approach has been already used in disciplines such as climate sciences, economics, and hydrology, also in combination with exploratory modeling (which in climate ensemble experiments are treated as an additional layer to the ensemble referred to as 'initial condition ensemble') (Tebaldi and Knutti 2007).

However, economics and hydrology have not been successful at integrating human and water systems (Pande and Sivapalan 2017). Conventional hydrologic (economic) models perceive pressures from human (water) systems, if considered at all, as external forcings. Where socioeconomic and hydrologic models interact in hydroeconomic models, responses to policy shocks or other *stimuli* are typically assessed using an external economic sub-model, which is subsequently integrated with the architecture of the hydrologic model through piecewise equations. This offers the advantage of a more straightforward and effective representation of causal relationships and interdependencies, while reducing computational costs since shocks do not require to be represented separately for each sub-model. Yet, such holistic models do not capture the interrelationships or two-way feedbacks between human and water systems that shape adaptive responses (Pande and Sivapalan 2017). As a result, the effects of policy- and climate-induced adaptation and feedback responses between socioeconomic, land surface, and water systems dynamics are still poorly understood.

There is a basic need to better understand the dynamics of complex adaptive human–water systems and to represent them in modeling tools that can be used to effectively inform policymakers. To this end, the transformative discipline of *socio-hydrology* has called for the development of integrated approaches that "explicitly account for the two-way feedbacks between human and water systems" (Sivapalan et al. 2014). Recent socio-hydrology-inspired science has explored feedback responses between human (typically water users) and water systems (Essenfelder et al. 2018). In parallel, economics has also developed new tools to explore feedback responses in complex human–human systems, notably between micro- and macroeconomic systems (Parrado et al. 2019). These contributions run standard models at each system level independently in *modules*, which are defined as specialized, self-contained mathematical elements that process information and generate outputs and connect them through sets of *protocols*, which are defined as rules designed to manage interrelationships (e.g. two-way feedbacks) between systems' modules (Csete and Doyle 2002). Modularity offers potentially higher detail in the representation of each system, which can be independently developed and adjusted. This makes possible the addition of non-linearity to each element of the system, so that surprises are not so surprising and can be adequately understood, and their repercussions transferred from one system to another.

While holistic models that use differential equations to capture as many systems as possible in comprehensive numerical models have significant practical value and continuing increases in computational power means, they can be systematically upgraded and adjusted to more accurately represent observed responses in human–water systems, it is reasonable to say that "we typically gain some understanding of a complex system by relating its behavior to that of other, especially simpler, systems" (Held 2005). It is through hierarchies of systems of increasing complexity, amenable to experimental manipulation that experimental sciences such as biology have made steady progress in, e.g., deciphering the human genome. Recently, climate research has put a stronger emphasis on model hierarchies as a means to link the complexity of high-end holistic simulations with a deeper understanding of the processes at work provided by conceptual models, so to discover previously unaccounted futures and explore their implied consequences. Analogously, to the extent that we can divide complex human–water systems into components that can be tested and developed in isolation, a hierarchy of human and water systems would make possible a more comprehensive understanding of the relevant processes involved through the use of conceptual models that capture their essence, and of the interrelationships among them through layers of feedback protocols (Csete and Doyle 2002).

I argue that recent advances in the construction of protocol-based modular frameworks provide the backbone for the development of interdisciplinary modeling hierarchies that connect multiple systems through two-way feedbacks (*multi-system hierarchy*). Each module within the hierarchy can be populated with multiple models (*multi-model ensemble*) and combined with scenario discovery techniques that explore scenario uncertainty through varying initial states and forcings (e.g. climate change scenarios, policy scenarios). The result is a large database of simulations in which each simulation represents the economic and environmental performance under one specific scenario and modeling setting. This information can be used to identify futures where proposed policies meet or miss their objectives, explore potential tipping points, and inform the development of robust policies that show a satisfactory performance under most conceivable futures.

Navigating Deep Uncertainty Type 2: Measuring and Understanding Transaction Costs to Avoid Techno-Institutional Lock-In

Assume the complete set of future outcomes in a system is R_A, where outcomes represent an event plus the policy response to that event. Through modeling we can reveal a fraction of the complete set $(R_A - r_\text{A})$, where $r_\text{A} = (\varepsilon_{1\text{A}} + \varepsilon_{2\text{A}})$, ε_1 is modeling limitations and ε_2 represents unawareness (the consequence of a priori unknowns). In coupled modeling frameworks, the second model or group of models in the hierarchy will then begin searching the repercussions of the feasible set $(R_A - r_\text{A})$ in a related system B and assess relevant feedbacks. Due to model limitations and unawareness, the coupled modeling framework will yield an incomplete set of future outcomes $(R - r_A * r_B)$, where R is the complete set of future outcomes in the coupled system and $r_B = (\varepsilon_{1B} + \varepsilon_{2B})$. Note again that by adding systems to the ensemble, modeling limitations and unawareness at each system level compound, increasing the range of possible future outcomes that we are unable to foresee. We can explore ways to limit the impact of ε_1 by adding and better representing models and scenarios across systems. This is indeed the objective of combining modular hierarchies with exploratory modeling and ensemble experiments. *However, ε_2 will persist until empirically revealed*. This is deep uncertainty type 2.

Deep uncertainty type 2 is the consequence of "limits in the knowledge base, chaotic dynamics, future actions by decision-makers, inherent randomness, non-stationarity and changes in societal perspectives and preferences over time", including stakeholders' preferences and their assessment of policies (Walker et al. 2003). Under deep uncertainty type 2, the only thing we know is that we do not know. Future predictions are "impossible", and society finds itself exposed to surprises, some of them potentially catastrophic ("black swans") (Taleb 2008). The natural question that follows is what can be done where the only thing we know is that we do not know. Critically, deep uncertainty type 2 is not an extreme on the scale of uncertainty—that place is reserved to *total ignorance* (Walker et al. 2003). Knowing we do not know gives us a valuable piece of information and allows us to plan in advance.

In addressing deep uncertainty type 2, the challenge is to strengthen **adaptive robustness** (Garrick 2015). Adaptive robustness involves the removal of techno-institutional barriers that constrain our ability to take corrective action so that incumbent policies can be replaced by superior alternatives as new information on possible futures is made available through the occurrence of surprises. Measuring and understanding techno-institutional barriers require information on *public transaction costs*, the institutional and organizational investments required to arrange, monitor, and enforce a policy. Public transaction costs include: (1) administering, monitoring, contracting, and enforcing current policy arrangements (termed *static transaction costs*) and (2) periodically designing, enabling, implementing new and/or transitioning existing management arrangements (termed *transition costs*). Transaction cost investments are also affected by (3) previous policy or institutional choices, which may enhance or constrain future selections (termed *technological and institutional lock-in* costs) (Loch and Gregg 2018).

Since predictions of future transaction costs are impossible under deep uncertainty type 2, anticipating the emergence of adaptively robust institutions is challenging. Yet, past transaction costs can be used to draw valuable insights into the trends and future development of adaptively robust institutions. The concept of *adaptive efficiency* is particularly useful in this regard. Adaptive efficiency measures the capacity of institutions to achieve economic efficiency over the long term. As compared with the conventional neoclassical approach, which views institutions as static and exogenous constraints within which costs and benefits are assessed, adaptive efficiency aims to understand long-term trajectories of institutional economic performance in contexts of entrenched path dependencies, complexity, uncertainty, and feedback between policy reform and implementation. In other words, adaptively efficient institutions are those showing "capacity to solve evolving and complex dilemmas over long periods of time, in a context of uncertainty and periodic, often unforeseen, shocks" (Garrick 2015). Note that **the concept of adaptive efficiency mirrors that of adaptive robustness**: adaptive efficiency looks at past institutional performance to individuate those institutions that were successful and efficient in taking corrective action; and adaptive robustness aims to remove constraints to the institutional ability to take corrective action in the future, so that future institutions are adaptively efficient. While ex-post adaptive efficiency does not equate to adaptive robustness, it is reasonable to expect that institutions that have proven to be adaptively efficient over long periods of time are more likely to be adaptively robust in the future. After all, the best thing we can do to predict the future is to prognosticate from the past. Just like (paleo)climatic data series can help narrowing the equilibrium response of global surface temperature to alternative CO_2 concentrations in climate models, or past choices are used to reveal agent's preferences and predict future behavior in economic models, data on past techno-institutional performance over sufficiently long periods of time can give valuable information to assess whether we are investing in institutions that are adaptively robust.

Garrick (2015) associates adaptive efficiency with "three performance indicators: (1) how well the objective(s) have been met (i.e. effectiveness); (2) the average public transaction costs per unit of the met objective(s); and (3) total program budgets". For an adaptively efficient institutional complex, these three performance targets should be "increasing, decreasing and sufficient", respectively (Garrick 2015). Although these three indicators are empirically measurable, public transaction costs are typically excluded from performance assessments of water or other environmental policies. In fact, transaction costs remain "a black box concept" for researchers, who rarely progress beyond zero transaction costs ideals (Loch and Gregg 2018). Although recent research has monetized transaction costs of water policy reform in South Africa, USA, and Australia, **the empirical base on transaction costs of water policy reform elsewhere is virtually non-existent**. **Moreover, in those areas where transaction cost data are available, studies usually do not quantify them over time** (Loch and Gregg 2018). Yet, measuring and analyzing adaptive efficiency to understand and predict future institutional performance, and whether it leads toward path-dependent/adaptively robust trajectories, necessitates longitudinal data on transaction costs.

Developing and analyzing longitudinal transaction cost data is in itself a major breakthrough that will help us understand the emergence of path-dependent/adaptively robust trajectories; yet, the natural question that follows is: *what can we do if past institutional performance leads to path-dependent trajectories that constrain our ability to take corrective action?* Existing technologies and institutions can constrain the range of policies that can be adopted in the coming years or decades through institutional and technological lock-in. In the context of water resource management, lock-in refers to the inertia of conventional engineering-based policies due to the mutually reinforcing physical, economic, and social constraints that emerge from existing technologies and institutions. Techno-institutional lock-in dynamics are driven by path-dependent increasing returns to adopted technologies and institutions at different levels: scale economies (production costs per unit decrease as fixed costs spread over growing production), learning economies (costs fall and performance improves as specialized knowledge and skills accumulate through experience), adaptive expectations (increased confidence about quality, performance, and permanence), and network economies (systemic relations among institutions, technologies, infrastructures, suppliers, and users). Water resource management is particularly prone to lock-in of conventional engineering-based policies due to large capital investments and long infrastructure lifetimes. The combined interrelationships between technological systems and basins' institutional matrices typically result in a self-referential system whose value increases with the growth of the techno-institutional complex (Unruh 2000).

The question of how to overcome techno-institutional lock-in in water resources management has received increasing attention in recent years. While traditional neoclassical economics argues that even marginal efficiency improvements are sufficient to drive the adoption of superior policies, empirical studies show that the inertia created within a techno-institutional complex necessitates an order-of-magnitude improvement in economic performance to induce transition, through **exogenous "annealing forces" that give change momentum** (Unruh 2000). Such an improvement is unlikely to arise endogenously from the techno-institutional complex. Public institutions typically show patterns characterized by incremental change, rather than transformational, over long periods, while examples of technology-led transformational responses are very limited (see below). The endogenous dynamics of a techno-institutional complex tend to create and reinforce its own stability or equilibrium, potentially leading to a path-dependent process of technological and institutional co-evolution that creates barriers to the diffusion of new, transformational policies.

Conclusions and Recommendations

Along these pages, I have surveyed recent advances in (socio-)hydrology and (institutional) economics toward robust decision-making; identified gaps in the development and integration of this research; and suggested a way forward in the integration of data and methods from natural and social sciences, so to deliver a research framework that informs the adoption of robust water policies with higher expected economic and environmental performance. Three major recommendations for future scientific work and research emerge from the analysis:

The development of a flexible and interdisciplinary modular hierarchy for the development of multi-system ensembles that incorporates and assesses the "two-way feedbacks" among modules, so to represent and understand the adaptive behavior of complex human–water systems.

The effort to gather longitudinal transaction cost data to create a database that supports analysis (notably through econometrics) toward a more in-depth understanding of institutional performance and key drivers of adaptive robustness, including "annealing forces" that impulse change and break up from path-dependent trajectories.

The integration of stakeholders in the generation of methods and results, so to underpin the emergence of valuable science-policy synergies that strengthen research quality and help identify statically/dynamically and adaptively robust policies with higher expected economic and environmental performance.

The three innovative elements above provide the pillars for the development of an interdisciplinary, replicable, and scalable research framework that *quantifies*the broad socioeconomic and environmental implications of adaptation policies and the uncertainty involved in the process, so to allow stakeholders to explicitly trade-off incremental increases in static/dynamic and adaptive robustness with expected policy performance, including policy costs, benefits, and effectiveness.

Beyond its scientific merit, the research agenda above have the potential to comprehensively test and demonstrate the performance of alternative solutions to water-related challenges and support decision-making toward the adoption of robust adaptation policies with potential to contribute to water policy objectives. Application of the research framework above will provide new insights for water policy as well as for the broader sphere of sustainable development. Understanding the implications of adaptation in terms of water reallocation and rationing, and related uncertainties, is relevant for policymakers who have committed to the good ecological status of water bodies, and also in terms of policy planning of related economic sectors (e.g. agriculture, agro-industry, tourism) and overall sustainable development, as substantiated in SDG 6 (UN 2015). Policymakers in these spheres need to be aware of trade-offs and distributive implications of adaptation policies in the water sector and below and will benefit from the methodological and empirical insights provided by the research agenda above.

References

Csete MF, Doyle JC (2002) Reverse engineering of biological complexity. Science 295:1664–1669. https://doi.org/10.1126/science.1069981

Essenfelder AH, Pérez-Blanco CD, Mayer AS (2018) Rationalizing systems analysis for the evaluation of adaptation strategies in complex human-water systems. Earth's Futur. https://doi.org/10.1029/2018EF000826

Garrick DE (2015) Water allocation in rivers under pressure: water trading, transaction costs and transboundary governance in the Western US and Australia. Edward Elgar Publishing, Cheltenham

Held IM (2005) The gap between simulation and understanding in climate modeling. Bull Am Meteor Soc 86:1609–1614. https://doi.org/10.1175/BAMS-86-11-1609

Loch A, Gregg D (2018) Salinity management in the Murray-Darling basin: a transaction cost study. Water Resour Res 54:8813–8827. https://doi.org/10.1029/2018WR022912

Marchau VAWJ, Walker WE, Bloemen P, Popper SW (2019) Decision making under deep uncertainty: from theory to practice, 2019th edn. Springer, Cham

Pande S, Sivapalan M (2017) Progress in socio-hydrology: a meta-analysis of challenges and opportunities. Wires Water 4:1–18. https://doi.org/10.1002/wat2.1193

Parrado R, Pérez-Blanco CD, Gutiérrez-Martín C, Standardi G (2019) Micro-macro feedback links of agricultural water management: insights from a coupled iterative positive multi-attribute utility programming and computable general equilibrium model in a Mediterranean basin. J Hydrol 569:291–309. https://doi.org/10.1016/j.jhydrol.2018.12.009

Sivapalan M, Konar M, Srinivasan V, Chhatre A, Wutich A, Scott CA, Wescoat JL (2014) Socio-hydrology: use-inspired water sustainability science for the Anthropocene. Earth's Futur 2:225–230. https://doi.org/10.1002/2013EF000164

Taleb NN (2008) The black swan: the impact of the highly improbable, Edición: trade paperback. Penguin, London

Tebaldi C, Knutti R (2007) The use of the multi-model ensemble in probabilistic climate projections. Philos Trans R Soc a: Math Phys Eng Sci 365:2053–2075. https://doi.org/10.1098/rsta.2007.2076

Unruh GC (2000) Understanding carbon lock-in. Energy Policy 28:817–830. https://doi.org/10.1016/S0301-4215(00)00070-7

Walker WE, Harremoës P, Rotmans J, van der Sluijs JP, van Asselt MBA, Janssen P, von Krauss MPK (2003) Defining uncertainty: a conceptual basis for uncertainty management in model-based decision support. Integr Assess 4:5–17. https://doi.org/10.1076/iaij.4.1.5.16466

World Bank (2016) High and dry: climate change, water, and the economy (report), water global practice. World Bank, Washington D.C. (US)

World Resources Institute (2019) Aqueduct - global flood analyzer [WWW Document]. Aqueduct. http://floods.wri.org/#/. Accessed 27 Nov 2019

Chapter 21
Cascading Transitional Climate Risks in the Private Sector—Risks and Opportunities

Hans Sanderson and Thomas Stridsland

Abstract Adaptation to climate change poses two recognized significant types of risks to the private sector; (1) physical risks and (2) transitional risks. As markets respond to climate-related policies and shifting demands from customers and investors, opportunities as well as risks are presented. A very recent and important policy development is the European Green Deal suggesting the EU to reduce its emissions from 40 to 55% by 2030, and aiming to enable European countries to meet their Paris Agreement targets. The shift required for this transition highlights the challenges in terms of adapting business models and decision-making tools, while also providing opportunities for innovation and development in the private sector. In order to reach Paris Agreement goals, science-based targets need to be adopted to measure and manage emissions, specifically focussing on Scope 3 emissions embedded in the value chain in the private sector. Methods and guidances are considered, with the ultimate goal being a harmonized methodology to create a detailed emissions inventory and risk disclosure of a company's operations. It is suggested that Environmentally Extended Input–Output models initially be used as a screening tool, in order to identify emission dense sectors. Process-based LCA inventory data, collected through collaboration and transparency throughout the value chain, can then be applied to increase the resolution of the decision-making tool.

Keywords Private sector · Transitional risks · Scope 3 emissions

Introduction

Adaptation to climate change is the management of identified climate-related risks; hence, a prerequisite to cost-effective adaptation is a sound, robust and quantitative

H. Sanderson (✉) · T. Stridsland
Department of Environmental Science, iClimate, iCSC, Aarhus University, Frederiksborgvej 399, 4000 Roskilde, Denmark
e-mail: sanderson@envs.au.dk

T. Stridsland
e-mail: str@envs.au.dk

C. Kondrup et al. (eds.), *Climate Adaptation Modelling*, Springer Climate,
https://doi.org/10.1007/978-3-030-86211-4_21

assessment of the risks. There are several different types of risks associated with climate change, it is, therefore, useful to split these into two large recognized categories, namely; (1) transitional risks and (2) physical risks. This paper addresses transitional risks with a focus on how they affect the private sector. The aim of this paper is to point out some of the main challenges facing the private sector in terms of adapting their decision-making and business models to the current and future transitional climate risk while contributing to a low-carbon economy and future. Transitional risks are systemic societal responses to climate change induced by, e.g. governments, such as increasing CO_2 pricing and taxes, but also novel demands from, e.g. investors, customers and the marketplace. The Task Force on Climate-Related Risk Disclosure (TCFD) (2020) has outline three different types of transitional risks companies need to assess as part of their analysis to comply with the TCFD recommendations (Table 21.1).

It is important to note that transitional risks are not only negative. To the diligent and responsible company, transitional risks represent opportunities in the market, they can exploit via new market developments, improved reputation and branding. From a corporate point of view, the transitional risk climate change represents a significant disruption of the market, and they have to adapt their business model to the new paradigm. The European Green Deal (EU 2020a) is an example of a potentially significant transitional risk. One of the objectives of the Green Deal is to enable the European Community to deliver our greenhouse gas (GHG) emission reductions in accordance with the target of the Paris Agreement to keep the global

Table 21.1 TCFD identified transitional risks

Market and Technology Shifts
Policies and investments to deliver a low carbon emissions economy:
Reduced market demand for higher carbon products and commodities
Increased demand for energy efficient, lower carbon products and services
New technologies that disrupt markets
Reputation
Growing expectations for responsible conduct from stakeholders, including investors, lenders and consumers:
Opportunity to enhance reputation and brand value
Risk of loss of trust and confidence in management
Policy and Legal
An evolving patchwork of requirements at international, national and state level:
Increased input/operating costs for high carbon activities
Threats to securing license to operate for high-carbon activities
Emerging concern about liabilities

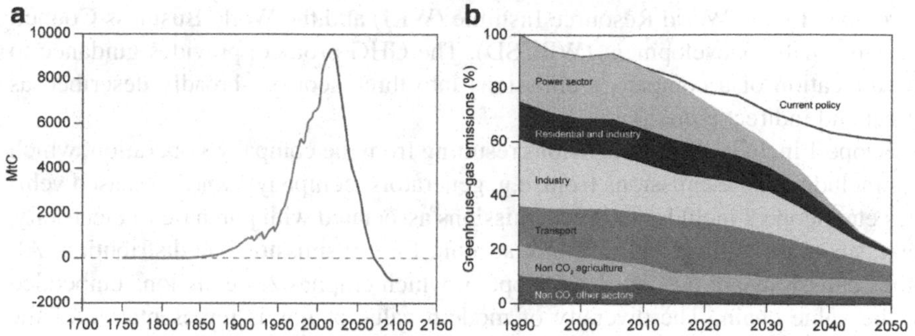

Fig. 21.1 **a** Past (blue) and future (red) global GHG emissions in MtC under the RCP 2.6 scenario; **b** Policy revisions needed for meeting GHG emission targets in the EU (EU 2020b)

temperature increase well below 2 °C and closer to 1.5 °C as represented by the IPCC RCP 2.6 scenario (Fig. 21.1a). It is clear that the sharp reduction targets will result in policy changes in several sectors, most of which are still on the drawing board (Fig. 21.1b). In other words, policy changes, GHG emission reduction demands and transition risks must be expected in the private sector. The industry (World Resource Institute (WRI); CDP; World Wildlife Foundation (WWF); United Nations Global Compact) in response have developed the science-based targets' initiative as a means for companies to comply with the Paris Agreement.

The sustainable finance action plan and the European taxonomy on sustainable finance (EU 2019) are vehicles for the Green Deal by steering the finance in a direction to economic investments in low-carbon assets. This is only possible if the market is transparent concerning the climate-related risks companies face (Sanderson et al. 2019). For most companies, the most eminent climate risk is their GHG emissions they need to assess, disclose, and explain how they will reduce these. These demands are developing rapidly as reflected in recent developments related to enhanced GHG emission reduction targets and pledges are considered as conditions in COVID-19 recovery support to private companies moving forward (ECB 2020). GHG accounting and report are in other words moving from an area of relatively lower importance within companies to a higher degree of concern and a risk that need to be assessed and managed in the same way as other significant risks companies need to manage and adaptation of business models.

Methods

As transitional climate risks become more existential to companies the methods to assess and report these must become more complete, clear and harmonized, than they are today as documented by Goldstein et al. (2019). Most companies today use the Greenhouse Gas Protocol (GHG Protocol) (GHG 2020a, b), which has been developed by the World Resource Institute (WRI) and the World Business Council for Sustainable Development (WBCSD). The GHG protocol provides guidance to the allocation of a company's emissions into three scopes—broadly described as direct and indirect emissions.

Scope 1 includes direct emissions resulting from the company's operation, which can include on-site emissions from, e.g. generators; company-owned or leased vehicles etc. Scope 2 includes indirect emissions associated with purchase of electricity, heat, steam or cooling, but does not account for transmission and distribution. All other emissions are described as Scope 3, which emphasizes emissions embedded in the value chain. The diversity of modern value chains is representative of the complexity a Scope 3 analysis can assume. Scope 3 encompasses upstream and downstream emissions and can include purchased services such as business travels to waste management. Note the risk of double accounting is present in Scope 3 analysis and needs to be avoided. The GHG protocol states that reporting Scopes 1 and 2 is mandatory, while Scope 3 can be considered optional and according to the WRI, some of the largest companies in the world only account and report their Scope 1 and 2 emissions. Most companies today only assess Scope 1 (direct emissions) and 2 (indirect emissions) and not the value chain-related emissions in Scope 3 (EU 2019). However, WRI has found, via a series of case studies, which the emissions along the value chain often represent a company's biggest GHG impacts, e.g. Kraft Foods found that value chain emissions comprise more than 90% of the company's total emissions (GHG 2020a, b). It is clear that inclusion of value chain-related emissions in Scope 3 is needed for an accurate and sound science-based target in compliance with the Paris Agreement.

Scope 3 emissions can be assessed and quantified using several different methods, as illustrated in Fig. 21.3 from Peters (2012). Environmentally Extended Input–Output models (EEIO) provide a top down assessment of all GHG emissions, including Scope 3 emissions. Such a process provides a comprehensive account of the value chain, establishing boundaries at the extraction of raw materials. This is done by coupling internationally available input/output (IO) tables with sector-wide emissions data, to determine emissions based on economic transactions. Despite EEIOs having advantageous coverage of upstream processes, combining sectors can result in sector aggregation errors, which has an effect on the resolution of the output (Acquaye et al. 2018). Alternatively, Life Cycle Assessments (LCA) provide an iterative, bottom-up approach, whereby emissions are quantified per functional unit, resulting in an emission factor (EF). EFs are defined by a detailed assessment of the environmental impact flows associated with the entire (cradle-to-grave) or part (cradle-to-X) of the product life cycle. These distinctions are significant for companies choosing a process-based method as it adds to complexities in defining Scope 3 boundaries. Valuable to this method is clearly defining the boundaries whereby emissions are accounted for, and upholding them throughout the organisation's analysis. Although this method is very detailed, double counting and omissions can occur as truncation errors when different product boundaries vary and are not accounted for (Lee and Ma 2013). Hybrid models incorporate the upstream coverage of EEIO and the downstream detail of LCA data in order to minimize aggregation and truncation

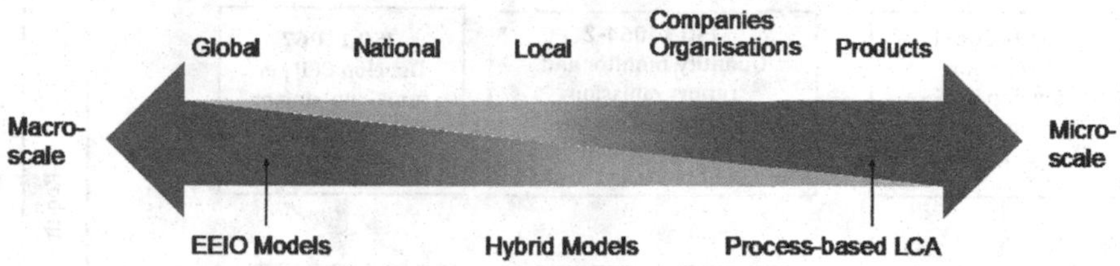

Fig. 21.2 Assessment continuum adapted from Peters et al. (2010)

errors. This requires an elaborate model and absorbs the complexities of both LCA and EEIO methods, resulting in seldom database updates, which can amplify the lessened uncertainties of both methods (Mattila et al. 2010).

Companies documenting their GHG emissions must choose their approach along the assessment continuum (Fig. 21.2), depending on which is most relevant to use as a tool for managing their emissions. It is not easy to orientate and find the right balance between resolution and do-ability, better accurate and pragmatic guidance is needed in this space. There are guidance documents for conducting an EEIO analysis in various multi-regional I/O tables, databases and models (Stadler et al. 2018; Yang et al. 2017), as well as process based and LCA—but there is still a lack of clear guidance and harmonization among the approaches, or combination of approaches, which hampers transparency in the reporting for investors.

The International Organization for Standardization (ISO) has also recently developed a series of standards for organizations to determine their GHG inventories as shown in Fig. 21.3, below.

The standards include quantification of value chain-related emissions in addition to the direct and indirect emissions (ISO 14064-1) and also quantification of GHG removal and sinks the organization may have (ISO 14064-2). The ISO standards are in other words quite comprehensive and complex, and probably beyond the current scope of most companies and organizations as documented by Goldstein et al. (2019). If the ISO standards are made mandatory, most companies and organizations would need to significantly upgrade their inventorying efforts.

Conclusions and Recommendations

As companies and organizations redefine their business models in compliance with their science-based target and the Paris Agreement, the broadening discussion on GHG emissions in the value chain becomes increasingly important. Up and down the value chain, collaboration is key, for these efforts to come together as a positive feedback mechanism. These will see business action cascade across their value chains and move, together, towards a low-carbon future in accordance with the Paris Agreement and the aspirations of the EU Climate Law and Green Deal proposed target of 55% reductions in the EU in 2030 compared with 1990 (Fig. 21.4).

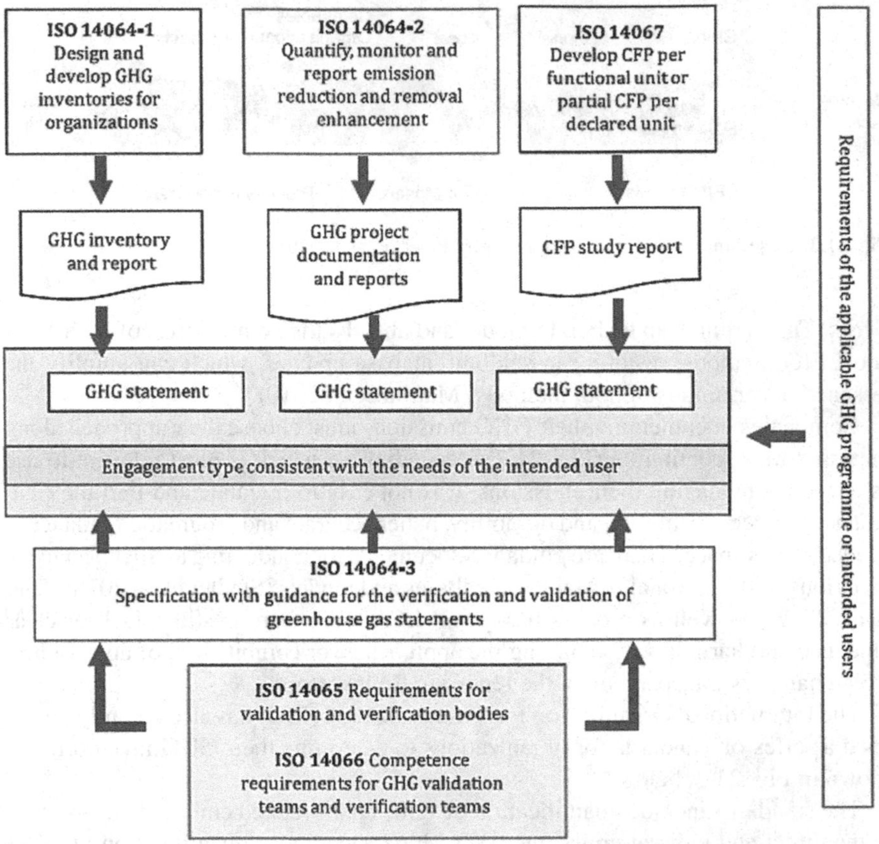

Fig. 21.3 ISO GHG inventory process for organizations

Crucial to this step is accurately assessing emissions such that innovation and development can occur. While EEIO can be used as a screening method to identify emission dense sectors, the nature of the analysis suggests the only action to reduce emissions is to reduce spending. Process-based approaches, however, provide the greatest resolution, which can act as a valuable tool for incorporating science-based targets and aligning with the Paris Agreement goals. By putting an emphasis on unit-based emissions, actors throughout the value chain are compelled to carry out their own assessments of their products, as a demand for emissions accountability grows. If ISO GHG standards are adopted as mandatory requirements, potentially conservative default values within Scope 3 can be defined. Doing so can incentivize companies and organisations to carry out Scope 3 analyses for a more accurate and less impactful representation of their company. While the disruption of climate-related transitional risks and associated complexities create challenges to companies addressing their Scope 3 emissions, it also creates huge opportunities for collaboration and innovation. Producers can develop innovative products and businesses that can open new markets

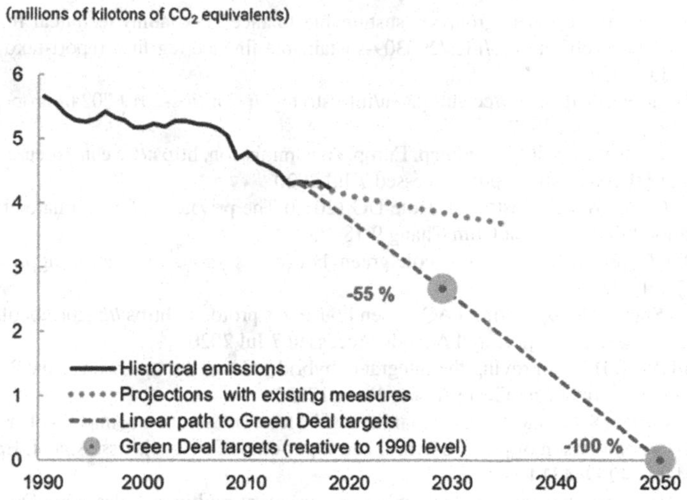

Fig. 21.4 EU total reduction targets

by adapting the way they do business in the value chain. GHG emissions can maybe in the future become an integrated part of value chain transactions much in the same way as money is today. Similarly, transparency of emissions accounting can be held to the same standard as economic transparency, e.g. beyond the nonfinancial disclosure in the stock exchanges. However, without the right standard method of GHG accounting, the concept of transparency in the market and green deal is difficult to ensure. This means that now is a key moment for defining a harmonized accounting method, as well as a pivotal time for companies to acknowledge their emissions, use the information to reduce them, and ensure their preparedness in the green transition. We recommend that value chain-related emission analysis and methods are included in the expertise areas of the Platform on Sustainable Finance (EU 2019) that need further clarification. There is a need for greater harmonization of practical guidelines for the assessment of value chain-related emissions to the industrial and financial sectors so they can accurately assess cascading transitional climate-related risks in the private sector (Sanderson 2021).

References

Acquaye A, Ibn-Mohammed T, Genovese A, Afrifa GA, Yamoah FA, Oppon E (2018) A quantitative model for environmentally sustainable supply chain performance measurement. Eur J Oper Res 269(1):188–205

ECB (2020) ECB on climate change. https://www.ecb.europa.eu/ecb/orga/climate/html/index.en.html. Accessed 7 Jul 2020

EU (2019) EU technical expert group on sustainable finance. Taxonomy technical report, June 2019. https://ec.europa.eu/info/files/200309-sustainable-finance-teg-final-report-taxonomy_en. Accessed 7 Jul 2020

EU (2020a) EU green deal. https://ec.europa.eu/info/strategy/priorities-2019-2024/european-green-deal_en. Accessed 7 Jul 2020

EU (2020b) Climate policy 2050 roadmap, European commission. https://ec.europa.eu/clima/sites/clima/files/2050_roadmap_en.pdf. Accessed 7 Jul 2020

Goldstein A, Turner WR, Galdstone J, Hole DG (2019) The private sector's climate change risk and adaptation blind spot. Nat Clim Chang 9:18–25

GHG (2020a) Green house gas protocol, green house gas protocol. https://ghgprotocol.org/. Accessed 7 Jul 2020

GHG (2020b) Standards supporting FAQ, green house gas protocol. https://ghgprotocol.org/sites/default/files/standards_supporting/FAQ.pdf. Accessed 7 Jul 2020

Lee CH, Ma HW (2013) Improving the integrated hybrid LCA in the upstream scope 3 emissions inventory analysis. Int J Life Cycle Assess 18(1):17–23

Mattila TJ, Pakarinen S, Sokka L (2010) Quantifying the total environmental impacts of an industrial symbiosis – a comparison of process hybrid and input-output life cycle assessment. Environ Sci Technol 44(11):4309–4314

Peters GP (2010) Carbon footprints and embodied carbon at multiple scales. Curr Opin Environ Sustain 2(4):245–250. https://doi.org/10.1016/J.COSUST.2010.05.004

Sanderson H, Irato DM, Cerezo NP et al (2019) How do climate risks affect corporations and how could they address these risks? SN Appl Sci 1:1720. https://doi.org/10.1007/s42452-019-1725-4

Sanderson H (2021) Who is responsible for embodied CO2? Climate. Preprints 2021, 2020120776. https://doi.org/10.20944/preprints202012.0776.v1

SBT (2020) Science-based targets. https://sciencebasedtargets.org/. Accessed 7 Jul 2020, and: change the chain: setting science-based targets for your value chain. https://sciencebasedtargets.org/2018/12/04/change-the-chain-setting-science-based-targets-for-your-value-chain/. Accessed 7 Jul 2020

Stadler K, Wood R, Bulavskaya T, Södersten CJ, Simas M, Schmidt S, Usubiaga A, Acosta-Fernández J, Kuenen J, Bruckner M, Giljum S, Lutter S, Merciai S, Schmidt JH, Theurl MC, Plutzar C, Kastner T, Eisenmenger N, Erb KH, Tukker A (2018) EXIOBASE 3: developing a time series of detailed environmentally extended multi-regional input-output tables. J Ind Ecol 22(3):502–515

TCFD (2020) Task force on climate-related risk disclosure. https://www.fsb-tcfd.org/. Accessed 7 Jul 2020

Yang Y, Ingwersen WW, Hawkins TR, Srocka M, Meyer DE (2017) USEEIO: a new and transparent united states environmentally-extended input-output model. J Clean Prod 158:308–318

Chapter 22
Climate Change Adaptation in Insurance

Marie Scholer and Pamela Schuermans

Abstract In this paper, we show three examples of how insurers can contribute to climate change adaptation, through insurers' underwriting and pricing practice. In the context of climate change, there is a clear need to go beyond traditional risk transfer products. Including risk reduction measures in an insurance product has the advantage of helping to better adapt to climate change by not only transferring the risk but by directly reducing avoidable damages when an event strikes, which as a result contributes to build a more resilient society.

Keywords Climate change · Adaptation · Insurance

Introduction

According to the 2020 World Economic Forum, more common extreme weather events could make insurance unaffordable or unavailable for individuals and businesses. Globally, the "catastrophe protection gap"—what should be insured but is not—reached US$280 billion (EUR€252 billion) in 2018. In Europe, only 35% of the losses from climate-related events are insured (EIOPA 2019).

There are limits to achieving a broad insurance penetration (Geneva Association 2019) and to climate change mitigation measures (European Commission 2020). Mitigation measures aim to avoid existing and new catastrophic events, whereas adaptation measures aim to limit the impact of these events. Adaptation as a means for anticipating the adverse effects of climate change and taking appropriate action to prevent or minimise the damage they can cause, or taking advantage of opportunities that may arise, plays an (increasingly) important role in ensuring society's resilience.

Insurance can contribute to limiting the impact (i.e. losses) of a catastrophic event, based on its knowledge and expertise in assessing (modelling, pricing) and pooling

M. Scholer (✉) · P. Schuermans
European Insurance and Occupational Pension Authority (EIOPA), Frankfurt, Germany
e-mail: Marie.Scholer@eiopa.europa.eu

P. Schuermans
e-mail: Pamela.Schuermans@eiopa.europa.eu

C. Kondrup et al. (eds.), *Climate Adaptation Modelling*, Springer Climate,
https://doi.org/10.1007/978-3-030-86211-4_22

risks. The European taxonomy for sustainable activities includes non-life insurance as an eligible activity, which substantially contributes to adaptation: insurance against climate-related hazards not only supports risk-sharing but also is working throughout the risk management cycle (identify, analyse, plan, implement and evaluate) and the disaster management cycle (prevent and protect, prepare, respond and recover) (TEG 2020).

Through adaptation measures, insurers can contribute to limiting the potentially systemic risks to society arising from climate change, such as economic consequences, in terms of welfare, including damage to the capital stock, sectoral productivity and changes in consumption (JRC 2020).

Adaptation measures that reduce insured risks will in the future define the sustainability of the insurance business model. In light of the potential impact and interconnectivity of risks posed by climate change innovative solutions for risk assessment, prevention and residual risk transfer among civil society, the market and public authorities are needed.

In this paper, we provide three examples where insurance could make a significant contribution to climate change adaptation, through insurers' underwriting and pricing practice. The price of insurance and the contractual terms and conditions under which insurance is being offered are generally strong signals about the risk even if commercial considerations also determine the price of insurance. By taking measures that influence the price, or the contract, insurers send a message on how they are managing the risks.

Incentivise Risk Reduction Measures in Property Insurance

The first example of how insurance could contribute to climate change adaptation is by providing incentives for risk reduction through premium discounts to policyholders who protect their property against natural catastrophes damages. We present an example of such a practice for floods prevention measures.

A number of prevention measures can be taken by policyholders to lower potential flood impacts (Hudson et al. 2016), such as limit the potential damage once the water has entered a building (known as wet flood-proofing) or attempt to prevent water from entering a building (known as dry flood-proofing). An illustrative example of damage reduction when dry flood-proofing has been implemented is shown in Fig. 22.1. Each of these measures will have different costs (Aerts 2018) and effectiveness.

One way insurers could incentivise risk reduction could be by giving premium discounts to policyholders, which would implement adaptation measures to minimise the risk. The insurer could even communicate this when a new customer asks for a flood insurance contract (i.e. provide premium with and without risk reduction measures). An example is shown in Table 22.1. We can clearly see that the cost of implementing the risk reduction measure is compensated by the lower premium. If policyholders would get access to this information, it might very likely increase the probability that they take risk reduction measures.

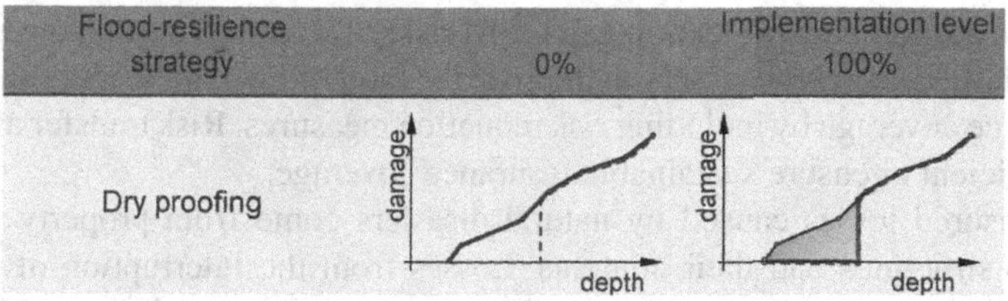

Fig. 22.1 Damage as a function of flood depth. Grey line: implementation level 0%; Dashed line: implementation level 100%; Light-blue area: reduced damage as the effect of flood-resilience technologies (Schinke et al. 2016)

Table 22.1 Based on the example of premium changes if risk reduction measures are considered (Kousky and Kunreuther 2014)

	Example	A zone [1] (USD)	V zone[2] (USD)
Premium	Premium for a house which is 3 feet below BFE[3]	4,000	18,550
Total cost: premium/year		**4,000**	**18,550**
Risk reduction measure	Cost to elevate to 1 foot above BFE	25,000 / annual loan (3%, 20 years) **1,660**	55,000 / annual loan (3%, 20 years) **3,660**
Premium	Premium 1 foot above BFE	**520** (more than 80% premium reduction)	**6,700** (more than 60% premium reduction)
Total cost: cost of risk reduction + premium/year		**2,180**	**10,360**

[1] A zone: is in the 1/100-year floodplain
[2] V zone: is in the 1/100-year floodplain in coastal areas
[3] Base Flood Elevation (BFE)—the estimated height of floodwaters during a 100-year period

In order to calculate the new premium based on different risk reduction measures, nat cat models would need to implement different vulnerability curves as shown in Fig. 22.1 (first using the one where the implementation level is 0% and then the one where implementation is 100%) and calculate the difference obtained in the estimated losses. These differences can then be reflected in the risk-based premium.

Promote Pro-active Management of Business Interruption Risks

Our second example of how insurance could make a significant contribution to climate change adaptation addresses the business interruption (BI) risks. Recent catastrophic events such as the Covid-19 pandemics or Hurricane Katrina have shown that there is a significant protection gap for BI risks. Taken the size that the BI losses can potentially take, it is necessary to improve the modelling of these risks and extend the insurance coverage by including risk reduction measures. Risk transfer alone will not be sufficient to ensure sustainable insurance coverage.

Most insured losses caused by natural disasters come from property damage: damage to structures and their contents. Losses from the interruption of business activities, however, can make up a significant portion of total losses and insured losses. Hurricane Katrina, for example, caused about $25 billion in insured commercial losses, of which 6–9 billion dollars has been attributed to BI (AIR 2008). There is also evidence that BI can be the most important variable in the survival of a business after a natural catastrophe.

In addition to BI losses arising from building and/or content damage, another source of losses after a natural catastrophe can also be non-damage BI (NDBI). Some businesses, such as aviation companies, might not be able to continue operating after a catastrophic event even if they were not directly impacted.

Standard basic BI insurance policies will cover an insured for losses arising from interruption to their business as a result of damage to insured property. BI insurance is often considered as an "annexe" to property insurance and no individual assessment of the BI risk is done. This may suffice in respect of a "normal" property loss, such as a fire or machinery breakdown. However, where there is damage to the wider area as a result of a natural catastrophe, this basic cover is often not broad enough. Considering the current property market sum insured for windstorms in Europe (see Fig. 22.2), the largest component of the total sum insured stems from buildings. The sum insured for BI represents only a small percentage for the commercial (industrial, agricultural and commercial) sector.

To assess BI risk, catastrophe models can be used. However, catastrophe models are usually well developed for assessing property damage but modelling of business interruption (BI) lags far behind (Rose and Huyck 2016). One reason is the crude nature of functional relationships in nat cat models that translate property damage into BI. Another reason is that estimating BI losses is more complicated because

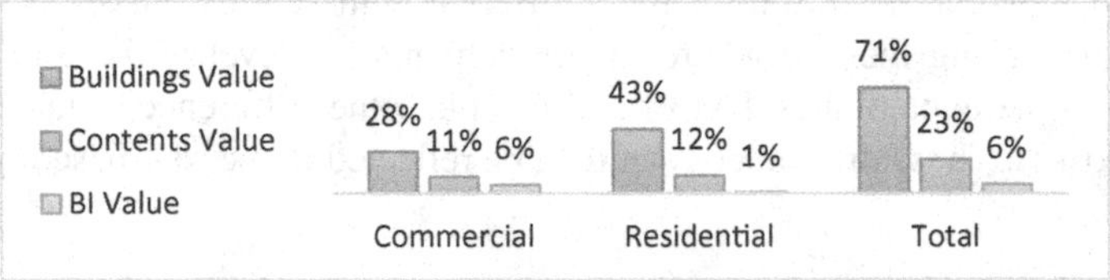

Fig. 22.2 Percentage of the total sum insured for building, content and BI for commercial (industrial, agricultural and commercial) and residential sectors for windstorms in Europe (data source: PERILS[1])

[1] https://www.perils.org/.

it depends greatly on public and private decisions during recovery with respect to resilience tactics that dampen losses.

BI insurance (for both BI and NDBI) is a key aspect to build a resilient society especially in the context of climate change adaptation. The survival of most businesses to natural catastrophes will depend on the extent to which an appropriate insurance policy has been chosen. Insurers need to offer adequate BI products by not only considering BI as an annexe to property insurance. In order for insurers to be able to offer adequate products, a clear BI risk assessment is needed to estimate the potential losses for proper risk-based premium calculation. Significant efforts should be put into the development of more accurate BI risk modelling. Insurers also have a clear role to promote the pro-active management of BI risks. They could, for example, offer premium rebates for undertaking contingency planning (e.g. plan to remove business activities from high hazard zones).

Improve Creditworthiness Through Adaptation Measures

Our third example shows that adaptation measures in property and NDBI insurance can contribute to reducing the risk from climate change in mortgage insurance and trade credit insurance (hereafter "credit insurance"). Additional adaptation measures from credit insurers can further contribute to climate change adaptation.

Creditworthiness, i.e. the ability to pay off one's debt, is central to mortgage and trade credit insurance. Mortgage insurance aims at paying off the outstanding debt in the event of the policyholder's death, disability, termination of employment or circumstances—specified in the policy—that may prevent the policyholder from earning income to service the debt. Trade credit insurance protects an insured from non-payment of commercial debt, i.e. it offers protection against the risk that a buyer defaults on a payment obligation.

Climate (change)-related costs are a source of credit risk: economic costs from physical and transition risks caused by climate change are ultimately borne by households and firms, affecting their cash flows and wealth, which are key determinants of their creditworthiness (Monnin 2018). Following a natural catastrophe, citizens may face the double burden of paying off a mortgage while also paying the reconstruction of their homes, potentially facing disaster-related unemployment. Businesses may face business interruption and/or physical damage to their property and, due to business interruption, create additional risks to the employment of individuals. All these risks are likely to lead to increased defaults on the payment for the purchase of goods and services or the servicing of a loan, leading to high incurred losses and pay-outs by credit insurers.

The impact of climate (change)-related events on mortgage insurance is well illustrated by the case study research by Corelogic on loan payment performance in Texas, after Hurricane Harvey hit the end of August 2017 (Corelogic 2018). In Hurricane Harvey FEMA (US Federal Emergency Management Agency) designated counties, properties estimated to have damage saw a 205% increase in 90+ day

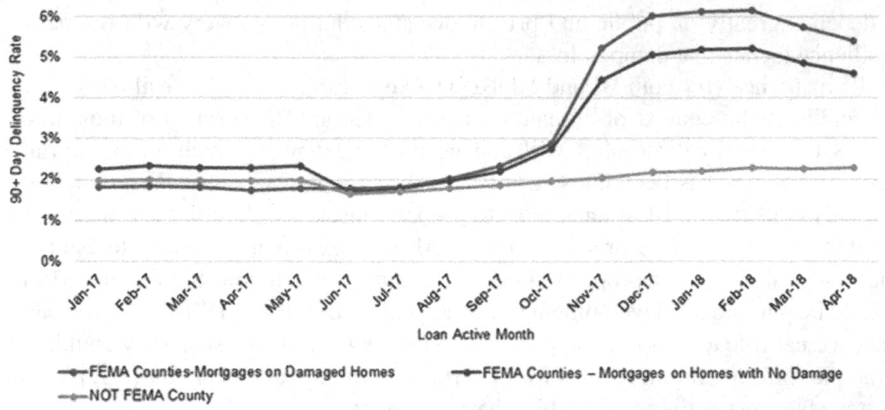

Fig. 22.3 90+ Day delinquency rate for TX . *Source* Corelogic 2018

mortgage delinquency, while properties estimated to have no damage saw a 167% increase in 90+ day mortgage delinquency when compared to delinquency rates just six months prior (Fig. 22.3).

As illustrated in our first and second examples, adaptive measures in property insurance and NDBI insurance can reduce insured risk. Adaptive measures to the property and business contingency planning can impact the cost for mortgage insurance or trade credit insurance, as these measures may contribute to reducing vulnerability to payment defaults arising from climate change. Applied to the case study of Corelogic: adaptation measures aimed at protecting the property against physical damages may lower the blue curve. Adaptation measures aimed at improving the business contingency planning to climate change scenarios may lower the green curve, across society.

Credit insurers can also themselves contribute to climate change adaptation in different ways: based on their risk assessment, credit insurers can support the insured in the buyer/sectoral risk assessment, to identify counterparties that may be particularly prone to default under climate (change)-related risks. Through an improved knowledge of the regional and global impact of climate change, credit insurers can also improve their country/region risk assessment.

Estimates of the damage and reconstruction cost of the property after a natural catastrophe can help improve the mortgage insurer's view of the risk of climate change to the mortgaged property, hence the relevance of looking at the underlying terms and conditions of the property insurance covering the mortgaged property. Based on the assessment of the possible longer-term financial burden on a policyholder following a disaster, a credit insurer could adapt the terms and conditions of the credit insurance (e.g. flexible premium payments, or temporarily lower deductibles).

Conclusions

In this paper, we have shown how an insurer could contribute to climate change adaptation, through insurers' underwriting and pricing practice. In the context of climate change, there is a clear need to go beyond traditional risk transfer products. Systematic consideration of how to integrate prevention measures to mitigate the insured risk should be the way forward. Including adaptation measures in an insurance product has the advantage to help to better adapt to climate change by not only transferring the risk (transferring the risk does not reduce the actual damage incurred) but by directly reducing avoidable damages when an event strikes. Examples were made of prevention measures for flood in property insurance, contingency planning for BI insurance and the impact of adaptation measures on credit insurance.

Climate change strengthens the need to have adequate risk transfer products, for example for business interruption as a significant protection gap exists for BI losses. Significant efforts should be put into the development of tools to identify protection gaps and in models to improve the quantification of risks and support risk-based pricing. Risk-based premiums are an important indicator of the evolving risk. Prevention measures aimed to reduce the insured risk can only be properly quantified and thus reflected in the premium if adequate models are available.

Going forward, if no adaptation measures are being taken, premia may become unaffordable or insurers' financial capacity may be lacking to cover for losses arising from climate change or other potentially systemic risks. Innovative risk solutions require relevant public and private stakeholders to share their data and exchange on different aspects of the risk assessment, prevention and transfer to ensure that risks remain insurable.

References

AIR (2008) Modelling business interruption losses. https://www.air-worldwide.com/publications/air-currents/modeling-business-interruption-losses/

Aerts J (2018) A review of cost estimates for flood adaptation. Water (Switzerland) 10(11):1–33. [1646]

Corelogic (2018) The impact of natural catastrophe on mortgage delinquency. https://www.corelogic.com/blog/2018/09/the-impact-of-natural-catastrophe-on-mortgage-delinquency.aspx

EIOPA (2019) EIOPA staff discussion paper - protection gap for natural catastrophes. https://www.eiopa.europa.eu/content/discussion-paper-protection-gap-natural-catastrophes_en

European Commission (2020) Adaptation to climate change. Blueprint for a new, more ambitious EU strategy. https://ec.europa.eu/clima/sites/clima/files/consultations/docs/0037/blueprint_en.pdf

Geneva Association (2019) Underinsurance in mature economies. Reasons and remedies. https://www.genevaassociation.org/sites/default/files/research-topics-document-type/pdf_public/underinsurance_in_mature_economies_web.pdf

Hudson P et al (2016) Incentivising flood risk adaptation through risk based insurance premiums: Trade-offs between affordability and risk reduction. Ecol Econ 125:1–13

Joint Research Centre (JRC) (2020) Climate change impacts and adaptation in Europe, JRC PESETA IV final report. https://ec.europa.eu/jrc/sites/jrcsh/files/pesetaiv_summary_final_report.pdf

Kousky C, Kunreuther H (2014) Addressing affordability in the national flood insurance program. J Extrem Events

Monnin P (2018) Integrating climate risks into credit risk assessment current methodologies and the case of central banks corporate bond purchases. CEP discussion note. https://www.cepweb.org/integrating-climate-risks-into-credit-risk-assessment-current-methodologies-and-the-case-of-central-banks-corporate-bond-purchases

Rose A, Huyck CK (2016) Improving catastrophe modelling for business interruption insurance needs. Risk Anal 36:1896–1915

Schinke R et al (2016) Analysing the effects of flood-resilience technologies in urban areas using a synthetic model approach. ISPRS Int J Geo-Inf 5(11):202

Technical Expert Group on Sustainable Finance (TEG) (2020) Taxonomy: final report of the technical expert group on sustainable finance. https://ec.europa.eu/info/files/200309-sustainable-finance-teg-final-report-taxonomy_en

Chapter 23
Climate Change Adaptation and Societal Transformation: What Are the Public Health Challenges?

Virginia Murray and Tim Chadborn

Abstract Behavioural change with societal transformation has been the key processes whereby hand and respiratory hygiene, social distancing and self-isolation that citizens across the world have been asked to implement to respond to the global COVID-19 pandemic. Is it possible to use such societal transformation approaches to change our behaviour for climate change adaptation? The European Commission (EC) funded research and innovation programmes that will be launched from 2021 will mobilise investment and EC's wide efforts to achieve measurable and time-bound goals on issues that affect citizens' daily lives. These programmes are based around five missions, one of which is the Mission on Adaptation to climate change including societal transformation. This will provide an opportunity to build evidence-informed assessment and design of interventions and should use a systems approach to determine and deploy the most cost-effective mix of public health behaviour change policy options according to the Nuffield Intervention Ladder and the Behaviour Change Wheel. This will maximise the likelihood of delivering societal transformation actions through ambitious but realistic research and innovation activities to help deliver planetary health programmes for Europe more widely.

Keywords Climate change adaptation · Societal transformation · Planetary health · Health behavioural change models

The European Commission's Missions for Horizon Europe

As part of the EC's budget for 2021–2027, the Commission proposed on 7 June 2018 that the next EC research and innovation programme, Horizon Europe, should have a proposed budget of €100 billion (European Commission 2018). By addressing

V. Murray (✉)
FRCP, FRCPath, FFPH, FFOM, Head of Global Disaster Risk Reduction, UK Health Security Agency, London, UK
e-mail: Virginia.Murray@phe.gov.uk

T. Chadborn
Head of Behavioural Insights and Evaluation Lead, UK Department of Health and Social Care, London, UK
e-mail: Tim.Chadborn@dhsc.gov.uk

C. Kondrup et al. (eds.), *Climate Adaptation Modelling*, Springer Climate,
https://doi.org/10.1007/978-3-030-86211-4_23

important societal challenges, such as climate change, through ambitious but realistic research and innovation activities, the European Commission (EC) has chosen to make clear to citizens how they can make a real difference in their lives, and in wider society as a whole (European Commission 2019a, b).

Missions are one of the main novelties of Horizon Europe and they were designed to boost the impact of EC-funded research and innovation by mobilising investment and EC's wide efforts around measurable and time-bound goals around issues that affect citizens' daily lives. To achieve their ambition the EC has organised five missions with one addressing **adaptation to climate change including societal transformation.**

In addition, the European Commission report on 'Adaptation to Climate Change-Related Health Effects - Scientific Advice to Strengthen the Resilience of the European Health Sector in View of Climate Change' states that there is a need to seek and prioritise synergies with climate mitigation actions and disaster risk reduction (European Commission 2020). It goes on to state that embedding 'human health' as a key component of the EU adaptation strategy should be aligned with the Sustainable Development Goals and with the Sendai Framework for Disaster Risk Reduction. This report recommends that there is a need to use the entire mix of policy interventions available at the EC level to intensify adaptation efforts in general, and particularly, the integration of health into climate adaptation.

Societal Transformation

UNESCO reports that '*the world is undergoing important social transformations driven by the impact of globalization, global environmental change and economic and financial crises, resulting in growing inequalities, extreme poverty, exclusion and the denial of basic human rights. These transformations demonstrate the urge for innovative solutions conducive to universal values of peace, human dignity, gender equality and non-violence and non-discrimination. Young women and men, who are the most affected by these changes, are hence the principal key-actors of social transformations*' (UNESCO 2019).

The COVID-19 pandemic represents a massive global health crisis. It has required large-scale behaviour change with societal transformation. To make such changes, the pandemic has placed and continues to place significant psychological burdens on individuals to help align human behaviour with the recommendations of epidemiologists and public health experts.

To address climate change—the other great challenge of our generation—it is apparent that the challenges associated with the application with societal transformation and adaptation models needed further consideration if evidence-based policy and decision-making in public and private sectors were to be effective. As a member of the **Adaptation to climate change including societal transformation** Mission, we identified the need for research to deliver the societal transformation that will engage all and recommended doing this using health behavioural change models.

This paper addresses some of the considerations around climate change drivers, recognises that societal transformation could limit the global average temperature rise, reflects what systems thinking for interventions might be available and finally how research might deliver evidence for the suggested interventions.

The Climate Change Driver

The global climate crisis is an existential threat to the world as we know it. Without a radical abatement of greenhouse gases, global warming will reach and exceed 3–4° before the end of the century (European Commission 2020). Climate change has already made some weather and climate extremes more frequent and severe. In 2019, the Lancet Countdown reports that the life of every child born today will be profoundly affected by climate change, with populations around the world increasingly facing extremes of weather, food and water insecurity, changing patterns of infectious disease and a less certain future. Without accelerated intervention, this new era will come to define the health of people at every stage of their lives (Watts et al. 2019). The COVID-19 pandemic has taught a lesson about how closely environmental, societal and human health are connected.

A second path—which limits the global average temperature rise to 'well below 2 ºC'—is possible and would transform the health of a child born today for the better, throughout their lives. Placing health at the centre of this coming transition will yield enormous dividends for the public and the economy, with cleaner air, safer cities and healthier diets.

Bold new approaches to policymaking, research and business are needed to change course. An unprecedented challenge demands an unprecedented response. It will take the work of the 7.5 billion people currently alive to ensure that the health of a child born today is not defined by a changing climate.

Working Towards the 'Second Path' to Limit the Global Average Temperature Rise to 'Well Below 2 ºC'

By identifying the key strategies to invest in for adaptation to climate change with societal transformation is briefly summarised as planetary health processes such as:

Enhancing early warning for extreme weather events and their cascading and complex hazards and risks (behaviour change and resilience especially for people at high risk such as with chronic obstructive pulmonary disease).

Emphasising the value of active travel and public transit (reducing exposure and minimising the production of vehicle fumes).

Reducing outdoor (e.g. coal power station outputs) and household air pollution (e.g. wood-burning stoves).

Encouraging sustainable and healthy diets (e.g. reducing meat consumption).
Reducing the use of fossil fuels and plastics.
Building sustainable and healthy cities.

What Systems Thinking Approaches for Interventions are Available?

Public Health England and its partners use the latest behavioural science to tackle the problems of complexity, focusing on developing multidimensional approaches to support healthy behaviours (PHE 2019). The PHE Strategy 2020–2025 recognises that behaviour occurs in a system and that we need to address multiple factors that influence behaviour and the behaviours of multiple actors at different levels of the system (e.g. for wood burning: citizens; retailers and manufacturers of wood burners; retailers of alternative electric heaters etc.).

The Interventions Ladder of Bioethics (Nuffield 2007) and Behaviour Change Wheel (Michie et al. 2011) frameworks can help plan by describing the degree of possible government intervention (with the most directive at the top to the least directive at the bottom) and categorising the methods that can be used to intervene: legislation, regulations, fiscal measures, guidelines, service provision, communications, media networks and marketing, environmental/social planning (such as building cycle routes/supporting social movements). Using a combination of complementary methods, a planned and synergistic systems approach to deliver the desired goals efficiently and equitably can be considered, for example:

Eliminate Choice: e.g. *Legislate* to stop highly polluting vehicles entering urban areas.
Restrict Choice: e.g. *Regulate* with standards for approved wood burners in smokeless zones.
Guide choice: e.g. *Fiscal Measures* to increase the cost of fuel to shift purchases towards electric vehicles.
Guide choices by incentives: e.g. offering tax-breaks for the purchase of bicycles for travel to work.
Guide choices by changing the default policy: e.g. by encouraging more restaurants to provide vegetarian-only cuisines.
Enable choices: e.g. enable individuals to change their behaviour by building cycle lanes.
Provide information: e.g. campaigns by encouraging the reduced use of plastic carrier bags.
Do nothing or simply monitor the current situation.

Recognising that behaviour change is critical for adaptation to climate change, we must improve understanding of the key behaviours involved in driving change, the key influencing those behaviours and what mix of interventions works best to

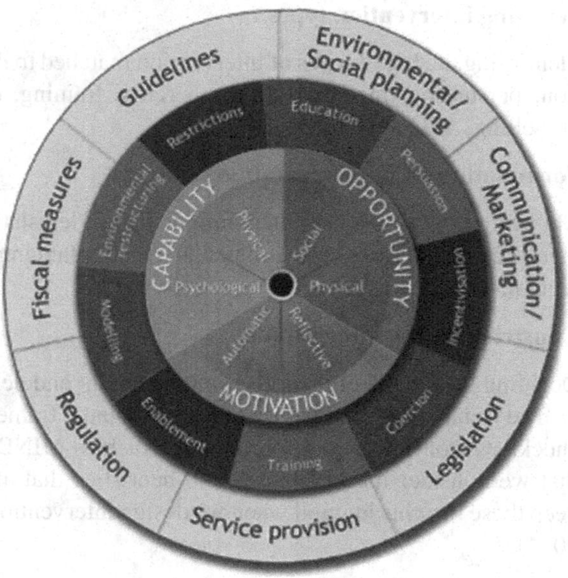

Fig. 23.1 Behaviour Change Wheel (Michie et al. 2011)

achieve impact. Most importantly, we need to use validated frameworks to ensure that our interventions are designed to target the key influences on behaviour based on evidence and theory and that we consider and utilise the full range of options available to us.

The Behaviour Change Wheel can help to address this by:
Developing behaviour change interventions from scratch,
Building on or modify existing interventions and choosing from existing or planned interventions (Fig. 23.1).

The Behaviour Change Wheel involves a number of processes to achieve and these include:

Assessment.

Evaluating the appropriateness of existing or proposed interventions in terms of the 'APEASE' criteria: **A**cceptability, **P**racticability, **E**ffectiveness, **A**ffordability, **S**ide-effects, and **E**quity (Annex 1).

Behaviour selection.

Identifying and selecting key behaviours to focus on in order to achieve policy objectives.

Capability, opportunity and/or motivation (COM-B) diagnosis.

What will most likely bring about the desired behaviour—in terms of changes in the target group's capability, opportunity and/or motivation to engage in/stop the behaviour.

Selecting intervention types.

Identifying the broad types of intervention matched to the COM-B diagnosis: education, persuasion, incentivization, coercion, training, environmental restructuring, modelling, and enablement.

Formulating an implementation strategy.

Choosing how to deliver interventions using: legislation, service provision, fiscal measures, guidelines, environmental/social planning, comms/marketing, and/or regulation.

Constructing the intervention.

Deciding the details of the intervention content and delivery.

The Behaviour Change Wheel and associated frameworks can be used alongside checklists from behavioural economics such as MINDSPACE (Annex 2) to ensure that we consider the key biases and heuristics that affect people's behaviour and keep these insights in mind when we design interventions and policy (Cabinet Office 2014).

How Research Might Deliver Evidence for the Suggested Interventions

To maximise the chances of success, it will be important to adopt a research-focused and systems-thinking approach such as the paper on building the case for systems thinking about climate change and mental health (Berry et al. 2018). Systems mapping with stakeholders and systematic reviews of the quantitative and qualitative evidence will collate what we already know about the behaviours, their influences and who performs them in each part of the complex climate change system. It will also integrate the evidence of which interventions have worked, and not, in the past, to what extent, for whom, to what cost, and in what context (real-world and research interventions).

Systems mapping and evaluation also need to be applied to existing policies and interventions so that we develop a comprehensive understanding of what we are already doing, where are the gaps and opportunities and where to focus resources—as informed by the understanding about the behaviours above. This can identify the need for new interventions or the need to enhance the behaviour change content or delivery of existing interventions. For each intervention, it will be critical to engage in detailed monitoring and evaluation to assess the impact. For example, the Public Health England Heatwave Plan and its annual reviews is an example of such monitoring and evaluation (CCC 2018).

Conclusions and Recommendations

The COVID-19 pandemic represents a massive global health crisis. Behavioural change with societal transformation has been the key processes whereby control measures of effective hand and respiratory hygiene, social distancing and self-isolation that have been implemented by citizens across the world have been asked to implement to respond to the global COVID-19 pandemic. Is it possible to use such societal transformation approaches to change our behaviour for climate change adaptation?

The recommendations from the Mission on Adaptation to climate change including societal transformation provide an opportunity to build evidence-informed assessment and design of interventions. By using a systems approach to determine and deploy the most cost-effective mix of policy options according to the intervention ladder and the Behaviour Change Wheel will assist in delivering societal transformation actions through ambitious but realistic research and innovation activities. These EC-funded research and innovation programmes that will be launched from 2021 will mobilise investment and EC's wide efforts around measurable and time-bound goals around issues that affect citizens' daily lives.

This approach should help to lead us to a future where the planetary health processes that are so vital for the future and summarised in Fig. 23.2 are the new normal.

Fig. 23.2 Planetary Health cartoon Wellcome Trust (2019)

Annexe 1 MINDSPACE is a Checklist of Influences on Our Behaviour for Use When Making Policy (Cabinet Office 2010)

Messenger: we are heavily influenced by who communicates information.

Incentives: our responses to incentives are shaped by predictable mental shortcuts such as strongly avoiding losses.

Norms: we are strongly influenced by what others do.

Defaults: we 'go with the flow' of pre-set options.

Salience: our attention is drawn to what is novel and seems relevant to us.

Priming: our acts are often influenced by sub-conscious cues.

Affect: our emotional associations can powerfully shape our actions.

Commitments: we seek to be consistent with our public promises and reciprocate acts.

Ego: we act in ways that make us feel better about ourselves.

Annexe 2 The APEASE Criteria for Assessing Interventions, Intervention Components and Ideas (Michie et al. 2014)

Acceptability	How far is it acceptable to key stakeholders? This includes the target group, potential funders, practitioners delivering the interventions and relevant community and commercial groups
Practicability	Can it be implemented at scale within the intended context, material and human resources? What would need to be done to ensure that the resources and personnel were in place, and is the intervention sustainable?
Effectiveness	How effective is the intervention in achieving the policy objective(s)? How far will it reach the intended target group and how large an effect will it have on those who are reached?
Affordability	How far can it be afforded when delivered at the scale intended? Can the necessary budget be found for it? Will it provide a good return on investment?
Side-effects	What are the chances that it will lead to unintended adverse or beneficial outcomes?
Equity	How far will it increase or decrease differences between advantaged and disadvantaged sectors of society?

References

Berry HL, Waite TD, Dear KBG, Capon AG, Murray V (2018) The case for systems thinking about climate change and mental health Nature Climate Change 8:282–290. https://doi.org/10.1038/s41558-018-0102-4)

Cabinet Office and Institute of Government Policy (2010) MINDSPACE Influencing behaviour through public policy. https://www.instituteforgovernment.org.uk/sites/default/files/publications/MINDSPACE.pdf. Accessed 6 Dec 2019

Directorate-General for Research and Innovation (European Commission), Group of Chief Scientific Advisors (European Commission 2020) Adaptation to health effects of climate change in Europe. https://op.europa.eu/en/web/eu-law-and-publications/publication-detail/-/publication/e885e150-c258-11ea-b3a4-01aa75ed71a1 accessed on 22 July 2020

European Commission (2018) EU budget: Commission proposes most ambitious Research and Innovation programme yet. https://ec.europa.eu/commission/presscorner/detail/en/IP_18_4041. Accessed 22 July 2020

European Commission (2019a) Commission launches work on major research and innovation missions for cancer, climate, oceans and soil. https://ec.europa.eu/info/news/commission-launches-work-major-research-and-innovation-missions-cancer-climate-oceans-and-soil-2019-jul-04_en. Accessed 22 July 2020

European Commission (2019b) Mission area: Adaptation to climate change including societal transformation. https://ec.europa.eu/info/horizon-europe-next-research-and-innovation-framework-programme/mission-area-adaptation-climate-change-including-societal-transformation_en. Accessed 22 July 2020

European Commission (2020) Accelerating the transition to a climate prepared and resilient Europe - Interim report of the mission board for adaptation to climate change, including societal transformation. https://op.europa.eu/en/web/eu-law-and-publications/publication-detail/-/publication/1d5234b9-b68a-11ea-bb7a-01aa75ed71a1. Accessed 22 July 2020

Michie S, van Stralen MM, West R (2011) The behaviour change wheel: a new method for characterising and designing behaviour change interventions. Implementation Sci 6:42. https://doi.org/10.1186/1748-5908-6-42

Michie S, Atkins L, West R (2014) The APEASE criteria for designing and evaluating interventions. In: The Behaviour Change Wheel: A Guide to Designing Interventions. London: Silverback Publishing. http://www.behaviourchangewheel.com/. Accessed 8 Dec 2019

Nuffield Council on Bioethics Public health: ethical issues 2007. https://nuffieldbioethics.org/assets/pdfs/Public-health-ethical-issues.pdf. Accessed 5 Dec 2019

Public Health England Strategy 2020–2025 (2019). https://assets.publishing.service.gov.uk/government/uploads/system/uploads/attachment_data/file/831562/PHE_Strategy_2020-25.pdf. Accessed 8 Dec 2019

UNESCO (2019) Social transformations. https://en.unesco.org/themes/social-transformations. Accessed 22 July 2020

UK Committee on Climate Change (CCC) (2018) Heat and Preventable Deaths in the Health and Social care System. https://www.theccc.org.uk/wp-content/uploads/2019/07/Outcomes-Heat-preventable-deaths-case-study.pdf. Accessed 8 Dec 2019

Watts N, Amann M, Arnell N et al (2019) The 2019 report of the Lancet Countdown on health and climate change. Lancet 394:1836–1878. https://www.thelancet.com/journals/lancet/article/PIIS0140-6736(19)32596-6/fulltext. Accessed 8 Dec 2019

Part IV
Adaptation Modelling and Policy Action

Francesco Bosello
Fondazione CMCC, Centro Euro-Mediterraneo Sui Cambiamenti Climatici and Department of Environmental Sciences, Informatics and Statistics, University of Venice Ca' Foscari, Venice, Italy
francesco.bosello@cmcc.it

Introduction

The EU Adaptation Strategy emphasised the need to "support better-informed decision making on climate adaptation at multiple governance levels". The new Adaptation Strategy of the EU, presented in March 2021, iterated the concept. It is based upon four pillars. It must be "smarter", "faster", "more systemic" and "stepping up international action". All the pillars, especially the first two, require access by decision-makers to the best available knowledge on the social and economic consequences of climate change and the cost and effectiveness of adaptation action. The knowledge must be useful or, said differently, of practical application within the decision process. This further calls for a form of translation or tailoring to the specific contexts and users. Especially in policymaking, this also means that the production and availability of information must meet the timing of the decision process that often is faster than that of science.

In a way, these needs are at the core of the development of adaptation modelling. Models are indeed tools that simplify complexity enabling better management and decision under uncertainty.

But where do adaptation models stand today? What are the barriers to their effective uptake in policy actions? Notwithstanding progress, at least three major gaps remain:

Many models still provide information with a spatial–temporal resolution which is too coarse compared with that of many adaptation policy decisions, particularly those at the urban, municipal level.

– Climate change impact, vulnerability and adaptation assessments can exceed the funding and technical capacity of smaller administrations. Moreover, the time needed to release such analyses is often too long compared to that of decision-making.

– There is not yet a common and consolidated practice in the communication of uncertainty. Current assessments do not always enable to disentangle uncertainty coming from the climate component, the social component, the models used and the parameterisation used.

– It finally emerges that most models cover climate change risk and impact assessment while a much more limited number of instruments relate explicitly to "identifying, assessing, implementing and monitoring" adaptation policies.

In the light of these gaps, some actions for improvements can be suggested. First, it is essential to increase the availability, reliability and accessibility of climate information with the "right" spatial resolution. Implementing some "front office" activity for producers of climate change data, primarily within the Copernicus Climate Change Service, can be useful in that. Similarly, the quantitative empirical basis on the cost and effectiveness of climate change adaptation has to be improved and made more accessible to decision-making. Particular attention has to be placed on local administration. This can be done, for instance, by promoting a systematic survey of existing quantitative evidence on costs and effectiveness of adaptation building on the results from many FP7, H2020 projects and research initiatives in this direction and strengthening the role of Climate Adapt, the EU portal on adaptation. It is also strategic to develop systematic guidance tools and case studies in the form of "rapid and light touch" analyses compared to what is currently available. Clearly, these can neither be applied to all decision-making problems nor substitute deeper investigations. Nonetheless, they can be valid policy supports in a screening phase of adaptation assessments, when adaptation decisions clearly have a "low" or "no"—regret nature, when more extensive analyses are simply not feasible. Finally, it is important to support the development of robust decision-making under uncertainty.

In the section, the contribution by Lourens M. Bower emphasises the limits of purely "technical" analyses of climate change risks for useful support of adaptation action. It is argued that understanding local capacities, governance and preference is the key to determine "when", "where" and "what" adaptation is feasible. This strongly calls for integrating top-down modelling analyses with bottom-up local experiences.

Marc Zebisch, Stefano Terzi, Massimiliano Pittore, Kathrin Renner and Stefan Schneiderbauer address the starting phases of the adaptation process. They present a conceptual framework for climate risk assessment enabling a "user-friendly" communication of complex cause–effect relationships in climate change impacts and risks and identify, accordingly, entry points for adaptation measures. This can provide a useful basis to policymaking for the selection of appropriate models, indicators or guide more qualitative, expert-based assessments.

Carlo Giupponi shows some possible solutions to the "uncertainty" challenge. Elaborating on a series of case studies, the contribution demonstrates the potential of multi-criteria analysis coupled with uncertainty analysis to guide practical adaptation action with sound and operational support.

Alexandre F. Fernandez and Frank Mc Govern close the section presenting the perspective of the Joint Programming Initiative "Connecting Climate Knowledge for Europe" (JPI Climate). This pan-European intergovernmental initiative is currently developing "JPI Climate Knowledge Hubs" and potentially a "European Facility for Climate Change" as concrete responses to the increasing need for active support in the implementation of national, European and international climate strategies and policies.

Chapter 24
The Roles of Climate Risk Dynamics and Adaptation Limits in Adaptation Assessment

Laurens M. Bouwer

Abstract The performance of adaptation measures depends on their robustness against various possible futures, with varying climate change impacts. Such impacts are driven by both climatic as well as non-climatic drivers. Risk dynamics are then important, as the avoided risk will determine the benefits of adaptation actions. It is argued that the integration of information on changing exposure and vulnerability is needed to make projections of future climate risk more realistic. In addition, many impact and vulnerability studies have used a top-down rather a technical approach. Whether adaptation action is feasible is determined by technical and physical possibilities on the ground, as well as local capacities, governance and preference. These determine the hard and soft limits of adaptation. Therefore, it is argued that the risk metrics outputs alone are not sufficient to predict adaptation outcomes, or predict where adaptation is feasible or not; they must be placed in the local context. Several of the current climate risk products would fall short of their promise to inform adaptation decision-making on the ground. Some steps are proposed to improve adaptation modelling in order to better incorporate these aspects.

Keywords Risk · Vulnerability · Exposure · Dynamics · Adaptation limit

Introduction

The assessment of adaptation measures is usually supported by model simulations on the performance of different measures under future conditions. Such simulations are again the basis for providing strategic directions, and often also cost and benefit estimates of different possible packages of adaptation measures.

The original version of this chapter was revised: This chapter has been updated with the correct figure 24.1. The correction to this chapter is available at https://doi.org/10.1007/978-3-030-86211-4_28

L. M. Bouwer (✉)
Climate Service Center Germany (GERICS), Helmholtz-Zentrum Hereon, Hamburg, Germany
e-mail: laurens.bouwer@hereon.de

C. Kondrup et al. (eds.), *Climate Adaptation Modelling*, Springer Climate,
https://doi.org/10.1007/978-3-030-86211-4_24

In this contribution, two topics are addressed that are relevant for adaptation modelling. These are the following:

Dynamics of risk and

Adaptation limits.

The first topic concerns the issue of capturing the dynamics of risk, with changes in climate, exposure and vulnerability, with the uncertainty of the latter two being at least of similar magnitude as the change in climatic hazards. The second topic concerns the hard and soft limits of adaptation that need to be investigated in order to inform decision-making for both adaptation and mitigation.

In this paper, some of the issues related to these two topics are discussed and hope to contribute to improving the performance and relevance of adaptation modelling, and eventually the take-up of results to achieve implementation of adaptation actions.

Dynamics of Risk

The risk concept has become the major basis for impact modelling and an essential part of assessing the performance of adaptation measures. This has been spurred by the developments of several vulnerability and impact models and associated studies over the past two decades. The SREX report of the Intergovernmental Panel on Climate Change (IPCC) underlined the need to integrate the three components of climate risk, which are: climate hazard, exposure and vulnerability (sensitivity). This report firmly established that Handmer et al. (2012):

Risk is determined not only by climate but to a very large extent by socio-economic circumstances, which are location-specific;

Risk is dynamic because of variability and changes in climate, adding to the (monotonous) changes and trends in socio-economic developments, and changes in human behaviour and (autonomous) adaptive responses to climate risk.

These dynamics need to be considered, in order to more reliably project how adaptation measures will perform over multi-decadal time horizons.

Exposure Change

Studies into past climatic impacts have highlighted that dynamics in exposure indeed play an important role in shaping changes in risk over time, including increased exposure of people and asset values driving up the losses from flooding and cyclone (wind) impacts. From past evidence, it is clear that socio-economic drivers until now have been dominant, and outweighed any climatic drivers. In the most comprehensive review to date Pielke (2021), based on 54 studies into a wide variety of extreme weather impacts, demonstrates that almost all studies show that losses have risen because of an increase in exposure.

Surprisingly, several studies attribute past losses solely to climatic changes when obviously increasing exposure has been the major driver of such costs (see, e.g., Coronese et al. 2019; Frame et al. 2020). Such studies do not correctly interpret the risk concept as laid out by IPCC. More importantly, they are not helpful for informing decision-making on risk management and adaptation, as important drivers are overlooked. Socio-economic developments amplify the impacts from changes in climate and are essential to include.

For the assessment of the performance of planned adaptation measures, it is thus important to integrate the future dynamics of exposure. In many places, especially urban centres, it is expected that population and wealth will increase, leading to increasing losses from climate hazards if no adaptation measures are taken (Bouwer 2013). Modellers have therefore resorted to integrate exposure scenarios, whereby socio-economic information is used to estimate possible developments in population, economy and resulting asset values. This enables the projection of risk, in combination with scenarios for changes in climatic hazards. For flooding, this is an approach that has now found wide implementation, mostly through (spatial) scenarios for population and economic growth in models for future flood risk (e.g. Vousdoukas et al. 2018; Dottori et al. 2018). For many other climatic hazards, however, socio-economics are not included, and risk is simply projected forward using the present socio-economic exposure and fixed vulnerability. The resulting risk analysis, with potential underestimates of future risk, is not a reliable basis for the evaluation of adaptation measures. Several adaptation measures could have higher benefits when such changes are considered.

Vulnerability Change

It has been argued that the temporal changes in vulnerability, or sensitivity, due to extreme weather events have also shaped past impacts, in addition to exposure changes. It is therefore important to understand such past changes in vulnerability, and project possible future developments out into the future in order to capture the range of possible future climate change impacts, and the performance of adaptation actions (e.g. Mechler and Bouwer 2015). However, there is less empirical evidence of changes in vulnerability, as such changes are difficult to capture. It has therefore often remained an underappreciated topic in climate impact modelling and projections (Bouwer 2013) and also in the assessment of adaptation.

Meanwhile, there are several studies pointing to the importance of including such vulnerability reductions. These include impacts from hydrological hazards, such as river floods (Kreibich et al. 2017), coastal floods (Bouwer and Jonkman 2018), as well as drought impacts (Kreibich et al. 2019). For instance, for coastal floods, it

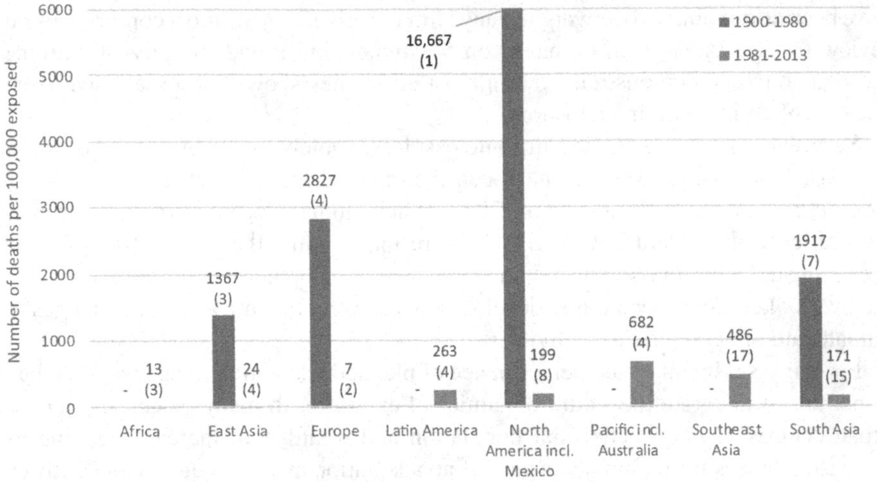

Fig. 24.1 Changes in average event mortality (number of deaths per 100,000 exposed population) for storm surge floods for different world regions between 1900 and 2013. The number of included events is given in brackets. Data from Bouwer and Jonkman (2018)

has been observed that not only morbidity (i.e. the number of casualties from these events) but also mortality (i.e. the relative death rate, or lethality) has decreased over time (see Fig. 24.1).

What is clear from Fig. 24.1, for instance, is that for most world regions, mortality rates have substantially declined, despite a strongly increasing coastal population and ongoing sea-level rise and land subsidence. For countries such as Bangladesh located in South Asia, this is remarkable. The strong decline is supposed to be the result of improved forecasting of cyclones, early warning, evacuation and shelters. In addition, in many areas improved coastal protection has resulted in less frequent flooding. The highest risk of deaths from storm surge flooding today is located in the Pacific and Southeast Asia.

Also for other hazards, such as extreme temperatures, it has been shown that substantial reductions in risk can be achieved by adaptation and preparedness actions, resulting in a reduction of vulnerability. For instance, Weisskopf et al. (2002) showed in a case study that a halving of deaths could be observed between consecutive heat waves, possibly as a result of substantial improvements in heatwave preparedness plans. At the global level the costs of weather-related hazards, as a share, are in fact going down (Formetta and Feyen 2019). Change in GDP here is a proxy for increasing exposed asset values. Similarly, deaths from such hazards are also declining compared to the total population (Formetta and Feyen 2019). It is also found that these rates have dropped more quickly for developing countries than for high-income countries, indicating the effects of progress on vulnerability reduction. However, a gap between the countries exists, as relative risks in low-income countries are still higher than in high-income countries (Formetta and Feyen 2019).

The problem with many of these studies is that they are empirical; i.e. they demonstrate some reduction in impacts that are unrelated to exposure or climatic changes, but cannot precisely attach these to causal changes in vulnerability (e.g. Weisskopf et al. 2002; Bouwer and Jonkman 2018; Kreibich et al. 2017).

In principle, there are two approaches for impact and adaptation modelling to account for such changes in vulnerability: one is to assume that vulnerability reduction is autonomous. This would be valid for several reductions in vulnerability that are related to emergency actions, such as preparedness actions at the household level. Such past trends, although not always underpinned by direct observational evidence, could be projected out into the future. Risk projections that include such trends can be used to assess adaptation options, thereby accounting for any risk reduction (increase) that is the result of any projected reduced (increased) vulnerability over time. Not including such substantial changes in vulnerability could overemphasise the effect of (additional) adaptation measures, and therefore the benefits of such investments.

The other approach is to assume that past adaptation actions are planned and can be directly observed. Such adaptation actions include the heightening of river dikes and coastal protection, improved forecasting and early warning systems, or improvements of water supply and the adjustments of crops. Such past actions can also be projected forward, and be taken as a baseline against which to compare additional or complementary adaptation actions.

Adaptation Limits

The interpretation of climate risk hinges on what the risk level means to the local population. It has been shown that climate impacts vary greatly over a given population, depending on their development status, income and other capacities. Therefore, some have argued that the risk metrics need to be adjusted to income levels before they can be correctly interpreted. For instance, Markhvida et al. (2020) show that poorer households in the San Francisco Bay Area suffer a much higher share of well-being losses compared to more affluent households. A single metric for impacts, such as monetary loss per capita, is therefore only partially useful, both in developing countries as well as developed countries.

Adjustments of climate risk metrics according to household income or other capacity information would help to highlight where the highest risks are located. Particularly high risks may indicate the need for additional adaptation measures to protect vulnerable populations and households. These high-risk levels are also indicative of limited capacities to deal with the impacts from climatic hazards, and we would argue that these levels may also be indicative of places where adaptation limits may be reached sooner.

Adaptation action can be limited by the local capacities to accommodate or reduce risk. While in many cases physical or technical options to reduce climate risks are available, and no hard adaptation limit is reached there, economic, social and governance constraints may lead to soft adaptation limits that are reached much earlier.

This is an area of investigation that has only recently received increasing attention (see, e.g., McNamara and Jackson 2019). These studies suggest that it is clear that there are limits to adaptation, and that the associated losses and damages need to be addressed. However, there is no practical framework yet to predict when limits are reached and when such losses would occur.

If top-down or local-scale modelling studies can show the physical-technical limits of certain measures and costs, then local and bottom-up studies are required to determine what capacities exist to actually implement such measures, and differentiate options depending on local preferences and possibilities. Importantly, such bottom-up studies can also show the limits for implementing such adaptation measures, both from a technical-physical perspective (hard limits), as well as soft limits (local capacities and preferences). This can be a starting point to indicate when and where limits may be reached and which losses beyond adaptation would occur.

Conclusions and Recommendations

The climate risk concept has now become a major basis for impact modelling, which in turn is an essential part of the assessment of the performance of proposed climate adaptation measures. However, this study argues that the dynamics of some autonomous risk processes are not yet sufficiently included in impact modelling. Also, the interpretation and actual meaning of risk assessment results both from an equity standpoint, as well as for assessing possible limits of adaptation and residual losses and damages, is currently insufficient. Several climate risk products would therefore fall short on their promise to inform adaptation decision-making on the ground.

Steps that would help to improve the potential of such products are:

The acknowledgement that past trends of exposure and vulnerability changes provide a baseline against which future risks should be compared

Better understand reductions in vulnerability that would add to any proposed adaptation measures

Tailor impact metrics from climate risk modelling to local situations, to account for equity issues and identify high-risk areas and hotspots where adaptation limits may be reached first

Use bottom-up studies to understand where local capacities, preference and governance could be hindering the implementation of required (technical) adaptation measures, and assess the risk of residual losses and damages.

Acknowledgements The author acknowledges the support through the Digital Earth project funded by the Helmholtz Association, funding code ZT-0025.

References

Bouwer LM (2013) Projections of future extreme weather losses under changes in climate and exposure. Risk Anal 33(5):915–930

Bouwer LM, Jonkman SN (2018) Global mortality from storm surges is decreasing. Environ Res Lett 13(1):014008

Coronese M, Lamperti F, Keller K, Chiaromonte F, Roventini A (2019) Evidence for sharp increase in the economic damages of extreme natural disasters. Proc Natl Acad Sci 116(43):21450–21455

Dottori F, Szewczyk W, Ciscar JC, Zhao F, Alfieri L, Hirabayashi Y, Bianchi A, Mongelli I, Frieler K, Betts RA, Feyen L (2018) Increased human and economic losses from river flooding with anthropogenic warming. Nat Clim Chang 8:781–786

Formetta G, Feyen L (2019) Empirical evidence of declining global vulnerability to climate-related hazards. Glob Environ Chang 57(6):101920

Frame DJ, Wehner MF, Noy I, Rosier SM (2020) The economic costs of Hurricane Harvey attributable to climate change. Clim Change 160:271–281

Handmer J, Honda Y, Kundzewicz ZW, Arnell N, Benito G, Hatfield J, Mohamed IF, Peduzzi P, Wu S, Sherstyukov B, Takahashi K, Yan Z (2012) Changes in impacts of climate extremes: human systems and ecosystems. In: Field CB et al (eds) Managing the risks of extreme events and disasters to advance climate change adaptation. Cambridge University Press, Cambridge, UK, pp 231–290

Kreibich H, Di Baldassarre G, Vorogushyn S, Aerts JCJH, Apel H, Aronica GT, Arnbjerg-Nielsen K, Bouwer LM, Bubeck P, Chinh DT, Caloiero T, Cortès M, Gain AK, Giampá V, Kuhlicke C, Kundzewicz ZW, Llasat MC, Mård J, Matczak P, Mazzoleni M, Molinari D, Dung NV, Petrucci O, Schröter K, Slager K, Thieken AH, Ward PJ, Merz B (2017) Adaptation to flood risk: results of international paired flood event studies. Earth's Future 5(10):953–965

Kreibich H, Blauhut V, Aerts JCJH, Bouwer LM, Van Lanen HAJ, Mejia A, Mens M, Van Loon AF (2019) How to improve attribution of changes in drought and flood impacts. Hydrol Sci J 64(1):1–18

Markhvida M, Walsh B, Hallegatte S, Baker J (2020) Quantification of disaster impacts through household well-being losses. Nat Sustain 3:538–547

McNamara KE, Jackson G (2019) Loss and damage: a review of the literature and directions for future research. Wiley Interdiscip Rev: Clim Chang 10:e564

Mechler R, Bouwer LM (2015) Understanding trends and projections of disaster losses and climate change: is vulnerability the missing link? Clim Change 133(1):23–35

Pielke R (s) Economic "normalization" of disaster losses 1998–2020: a literature review and assessment. Environ Hazards 20:93-111

Vousdoukas MI, Mentaschi L, Voukouvalas E, Bianchi A, Dottori F, Feyen L (2018) Climatic and socioeconomic controls of future coastal flood risk in Europe. Nat Clim Chang 8(9):776–780

Weisskopf MG, Anderson HA, Foldy S, Hanrahan LP, Blair K, Török TJ, Rumm PD (2002) Heat wave morbidity and mortality, Milwaukee, Wis, 1999 vs 1995: an improved response? Am J Public Health 92(5):830–833

Chapter 25
Climate Impact Chains—A Conceptual Modelling Approach for Climate Risk Assessment in the Context of Adaptation Planning

Marc Zebisch, Stefano Terzi, Massimiliano Pittore, Kathrin Renner, and Stefan Schneiderbauer

Abstract In this paper we present a conceptual framework for a climate risk assessment based on the so-called impact chains. The method follows a general assessment framework consistent with the IPCC AR5 concept on climate risk. This framework has been developed by Eurac Research within the context of various projects with the German Environment Agency and the German Gesellschaft für Internationale Zusammenarbeit (German Corporation for International Cooperation)—GIZ. It has been applied in almost twenty national climate risk assessments worldwide (e.g., Burundi, Bangladesh, Thailand, Vietnam, Madagascar) and has been perceived as (1) an appropriate means for risk analysis, (2) a useful tool for communication of complex cause-effect relationships in climate change impacts and risks, and (3) a great approach to identify entry points for adaptation measures. For an operational risk assessment, impact chains serve as a basis for the selection of appropriate models, indicators or guide more qualitative, expert-based assessments.

Keywords Climate risk assessment · Impact chains · Vulnerability · Adaptation

Introduction

A comprehensive and context-specific climate risk assessment (CRA) is a common and highly recommended step to prepare adaptation planning. Typically, climate risk assessments are built upon information on current climate extremes and scenarios of future climate, an analysis, which other underling factors and trends (ecosystem related, physical, technical, or socio-economic factors) are influencing climate risks

---

M. Zebisch (✉) · S. Terzi · M. Pittore · K. Renner · S. Schneiderbauer
Institute for Earth Observation, Eurac Research, Bolzano, Italy
e-mail: marc.zebisch@eurac.edu

S. Schneiderbauer
Institute for Environment and Human Security, United Nations University, GLOMOS Programme, Bolzano, Italy

C. Kondrup et al. (eds.), *Climate Adaptation Modelling*, Springer Climate,
https://doi.org/10.1007/978-3-030-86211-4_25

in various sectors, an assessment of potential impacts and risks associated with climate extremes and climate change on the respective sectors.

There is no standard way of how to conduct a CRA. However, an ISO-Norm (ISO/DIS 14091: Adaptation to climate change—Guidelines on vulnerability, impacts and risk assessment) is under preparation (ISO 2020) and containing a selection of suitable tools. National CRAs in Europe follow, in general, the sequence mentioned above, but without a common scheme. An EEA Report (EEA; ETC/CCA 2018) on national climate change vulnerability and risk assessments in Europe provides a good overview on the applied approaches.

Responding to the current lack of frameworks and guidelines for climate risk assessments, Eurac Research has created a conceptual framework called climate impact chains together with various partners. Impact chains were first developed for climate risk assessments in the European Alps (Schneiderbauer et al. 2013; Zebisch et al. 2014), and following applied for the 2nd national climate risk and vulnerability study for Germany (Buth et al. 2017). Finally, they were transformed into a set of guidelines for climate vulnerability and risk assessment for the German Gesellschaft für Internationale Zusammenarbeit—GIZ (Fritzsche et al. 2015; Zebisch et al. 2017, 2021; Hagenlocher et al. 2018). In the meanwhile, this framework has been applied in almost twenty national climate risk assessments worldwide (e.g., Burundi, Bangladesh, Thailand, Vietnam, Madagascar) and is proposed by the new ISO 14091 as one of the concepts for climate risk assessments in the context of adaptation planning. Moreover, a running JPI Climate project (www.unchain.no) is dealing with the application and improvement of the impact chain concept.

The terminology and concepts of impact chains are referring to the most recent climate risk concept of the IPCC, introduced within the IPCC AR5 in 2015 and further developed for the 2019 IPCC SROC report (Abram et al. 2019). According to the IPCC, climate risks are potential adverse consequences for human or ecological systems caused by climate extremes and climate change. A climate risk (e.g., drought damage in agriculture) results from the interactions between climate-related hazards (e.g., droughts) with exposure (e.g., agriculture land) and vulnerability (e.g., drought resistance of crops, presence or absence of irrigation) of the social-ecological systems. Adaptation strategies can reduce climate risks, mainly by reducing vulnerabilities, but also by reducing exposure or the climate-hazard itself (see Fig. 25.1).

Methodology—Impact Chains, Operationalisation and Adaptation Planning

Impact Chains—Conceptual Models of Climate Risks

Impact chains are conceptual models based on cause-effect chains that include all major factors and processes leading to specific climate risks in a specific context (e.g.,

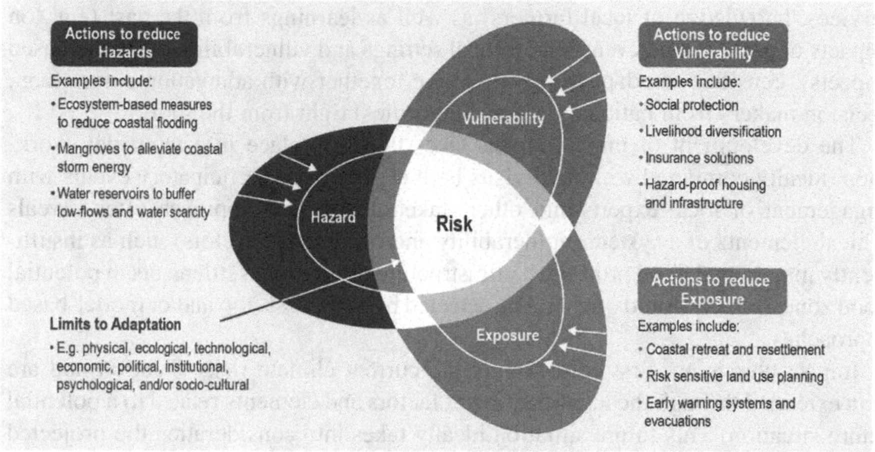

Fig. 25.1 Climate Risk as a function of hazard, vulnerability and exposure—from IPCC SROC (Abram et al. 2019)

regional and/or sectoral). Factors and processes are assigned to the risk components hazard, vulnerability or exposure, while cascading effects are considered as intermediate impacts (Fig. 25.2, left). Impact chains are usually developed in a participatory manner (Fig. 25.2, right) together with stakeholders and experts to create a commonly agreed picture of root-causes for climate risks in a specific context, allow the integration of local data and knowledge (e.g., data from national weather

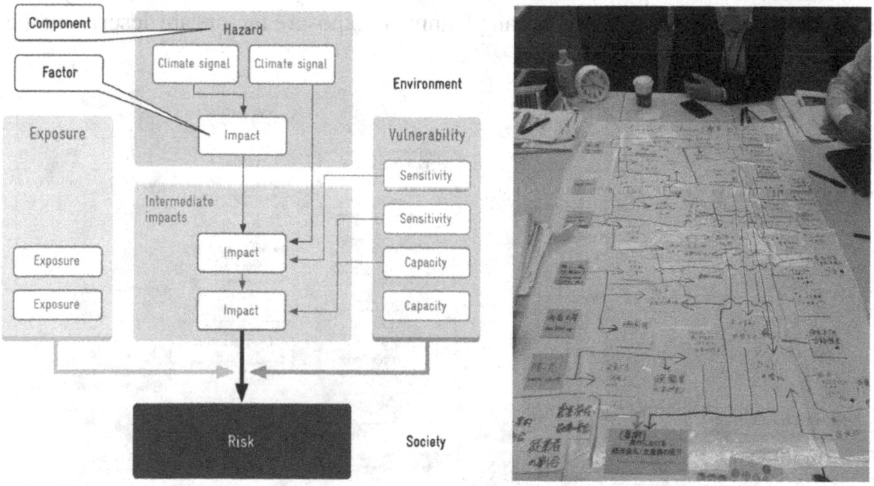

Fig. 25.2 Left: a generic climate impact chain with the risk components hazard, exposure and vulnerability as well as intermediate impacts; right: participatory approach to develop impact chains—case study from Japan (M. Zebisch)

services, knowledge of local farmers) as well as learnings from the past (e.g., on impacts of past climate events and critical settings and vulnerabilities that increased impacts), consider an adaptation perspective together with adaptation actors (e.g., decision makers from national or local authorities) right from the start.

The development of impact chains takes typically place in a multi-day workshop, ideally combined with field visits both organised as participatory events with engagement of local experts and other stakeholders. This approach often reveals critical elements of a system (vulnerability and/or exposure factors) such as insufficiently maintained flood protection infrastructure or informal settlements in potential flood zones, which would often not be detected by more desk-top and/or model-based approaches.

Impact chains are first built around the current climate risks situation and are than extended through the identification of factors and elements related to a potential future situation. This future situation ideally takes into consideration the projected future climate, but also potential future trends in exposure (e.g., urbanisation leading to a higher exposure values in cities) and vulnerabilities (e.g., an aging population leading to a higher vulnerability of population).

Figure 25.3 shows an example of a simplified impact chain for the 'risk of loss and damage due to floods' in a fictive river catchment in a mountainous environment in South-East Asia. While the hazard—impact—risk relation is often quite straightforward and tangible (e.g., heavy rain events are triggering floods and the related risk of loss and damage due to floods), vulnerability and exposure factors are predominantly complex, strongly context specific and partly intangible. Vulnerability factors include a wide range of issues spanning from natural, ecosystem related parameters and processes (such as a reduced natural retention capacity due to wetland degradation) to more technical parameters (e.g., absence of flood resistant houses) or capacity related parameters (e.g., lack of urban planning). Exposure factors are describing the

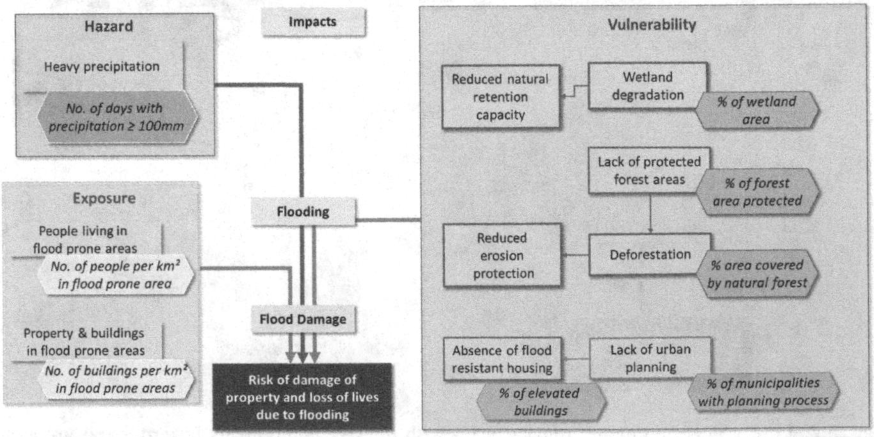

Fig. 25.3 Impact chain for the risk of loss and damage due to floods with single factors (squared boxes) and appropriate indicators for single factors (hexagonal boxes)

presence, number or density of assets and livelihoods potentially affected by a certain hazard.

Operationalisation—Models, Indicators, Qualitative Approaches

For an operational risk assessment, impact chains can serve as a backbone for the integration of various methods such as models, indicators or participatory data collection. Often, a combination of these methods is recommended. The general rule is "Take the best you get from and with the resources available". Importantly, the risk assessment should not be driven by data availability only. While for some components of a risk (e.g., information on current and future climate extremes) quantitative data is widely available or physical models exists (e.g., for hydrological risk), for other elements (vulnerability factors, more complex risks) an assessment can only be based on proxy indicators or expert assessments. We strongly recommend not to leave out any relevant risk or underlying process or factor just because respective data is not available. This approach is particularly important within the context of adaptation planning, where a qualitative understanding of specific vulnerabilities is often more important than quantitative, sometimes pseudo-precise, information on hazards or impacts.

The original approach of the Vulnerability Sourcebook (Fritzsche et al. 2015) proposes an indicator-based approach to quantify each risk component and finally the resulting risk. For each of the relevant factors from the impact chains, appropriate indicators are defined (see Fig. 25.3). Indicators are transformed into a common, normative value scheme (e.g., from 0 = optimal to 5 = critical), aggregated within risk components, and, finally, into a composite risk indicator for each specific risks. The advantage of such an indicator-based approach is its transparency and the ability to compare different sub-units (e.g., districts) or time slices. However, it requires a data-rich environment and involves a lot of decisions on normalisation, weighting and aggregation. Often, a more comprehensive, narrative aggregation of data and information collected along the factors and elements of an impact chain is a valid alternative.

A typical final result is a report with detailed description of each single climate risk consisting of a mix of narrative information, graphs and maps with information on which underlying factors lead to a specific climate risk in the specific context, which of them are particularly relevant and could be an entry point for adaptation, results from data analysis and models for the specific climate hazard, impact models and indicators, an overall risk assessment (high, medium, low) for the current situation and selected future time slices under different climate scenarios, if possible, on the level of spatial sub-units (e.g., districts or bio-geographical zones), spatial hotspots or specific settings that lead to a higher than average risk, uncertainties and confidence levels based on a qualitative assessment of different sources of uncertainty and disagreement.

How Impact Chains Support Adaptation Planning

Impact chains can foster the discussion on adaptation demand already in an early, qualitative stage of a risk assessment. Particularly, when developing the impact chains with stakeholders and discussing on vulnerability factors ("what makes your system vulnerable against a specific climate hazard?") specific sensitivities (e.g., drought-sensitive crop types) or the lack of technologies as well as capacities (e.g., the lack of an efficient irrigation system, the lack of a hazard zone planning or the lack of an integrated water resource management strategy) immediately pops up (examples of adaptation options for the fictive case in Fig. 25.4. Already during this phase, appropriate measures to tackle these deficits can be discussed and recorded. After the operational risk assessment, more details about critical settings or spatial hotspots are revealed, which allows for a more targeted discussion on adaptation options (e.g., where exactly an adaptation measure must be applied). A reverse discussion (which adaptation measures could be appropriate or have been applied in a similar context in other regions) may help to discover potential vulnerabilities that have been overlooked so far. Involving responsible stakeholders or experts for adaptation measures already in the risk assessment phase guarantees a higher commitment and a smooth and consistent transition from the risk assessment towards the adaptation planning phase.

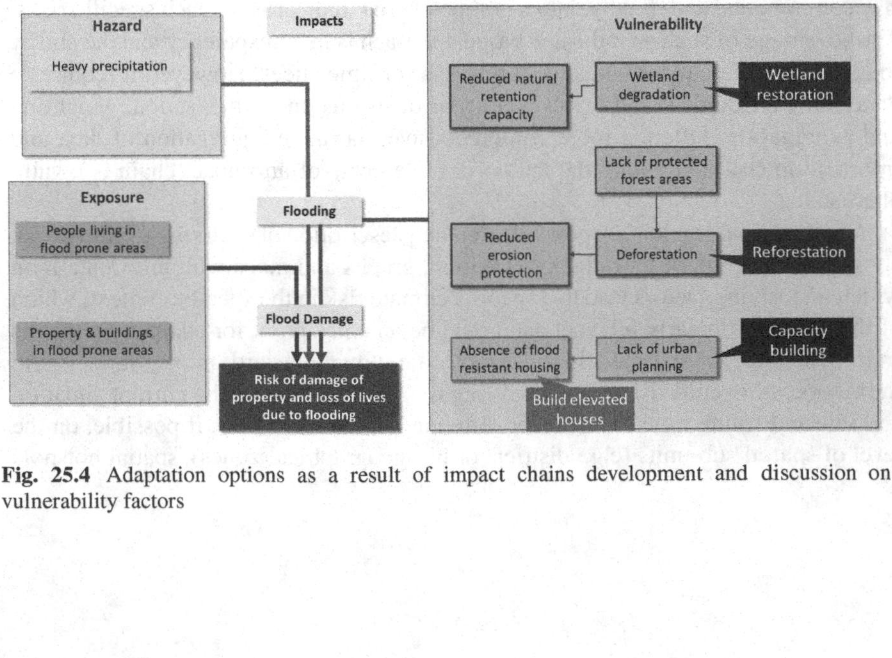

Fig. 25.4 Adaptation options as a result of impact chains development and discussion on vulnerability factors

Conclusion and Recommendations

In real-world applications, impact chains have been perceived as a very useful tool for analysis as well as for communication of complex cause-effect relationships in climate change impacts and risks. Particularly, the participatory approach has been highly appreciated in order to obtain a structured and agreed picture of context specific climate risks in-line with the IPCC AR5 climate risk concept. For a successful participatory process, it is important to gather relevant regional/local expertise for the selected climate risks and sectors (examples of such are agriculture, water management and tourism). At the national scale, this could include experts from national environmental ministries and agencies, line ministries and agencies, national statistical offices, national meteorological services, national universities as well as stakeholders from the private sector. All of them might also be data providers and, since data availability is often a bottleneck for the definition and population of indicators, data availability should be discussed with them early in the process.

Amongst the main advantages of impact chains is the possibility to highlight and describe relevant impacting mechanisms related to different hazards on complex and heterogeneous socio-ecological systems, also pinpointing causal relationships and relevant factors over a broad spectrum of vulnerability facets. Modelling such complex relationships requires a hybrid approach (quantitative/qualitative). While data on climate hazards (e.g., drought) is usually available for current and future climate, data on exposure (e.g., population density) or vulnerability is often available only for the current situation or completely missing. For physical factors or processes (e.g., land degradation), analysis from earth observation could be an appropriate approach to gather information on current status and trends. Expert based approaches are often the only solution to describe complex risks but have the disadvantage of a potential bias related to the selection of experts. A challenge related to this hybrid approach is validation. While climate models and some impact models can be validated for the current situation, indicator-based and expert-based approaches are often more hypothesis driven. Options for validation of hybrid and hypothesis driven methods are an interesting field for future research. Another question for future research is to which extend a climate risk assessment could integrate elements of a classical Disaster Risk approach by providing quantitative estimates of expected loss and damage and their likelihood.

Finally, every risk assessment is in the end a value-based approach. Risk depends on priorities, values, targets and normative settings. Therefore, in each risk assessment a discussion on which effects of climate extremes and climate change would fundamentally threaten a system ("high risk") and which effects might be tolerable or are within the coping capacity of a system ("low risk") should be conducted.

References

Abram N et al (2019) Framing and context of the report. In: IPCC special report on the ocean and cryosphere in a changing climate', IPCC

Buth M, Kahlenborn W, Greiving S, Fleischhauer M, Zebisch M, Schneiderbauer S, Schauser I (2017) Guidelines for climate impact and vulnerability assessments. Dessau

EEA; ETC/CCA (2018) National climate change vulnerability and risk assessments in Europe, EEA Report No 1/2018. Copenhagen

Fritzsche K, Schneiderbauer S, Bubeck P, Kienberger S, Buth M, Zebisch M, Kahlenborn W (2015) The vulnerability sourcebook. Edited by GIZ - Gesellschaft für internationale Zusammenarbeit. Eschborn. https://www.adaptationcommunity.net/?wpfb_dl=203

Hagenlocher M, Schneiderbauer S, Sebesvari Z, Bertram M, Renner K, Renaud F, Wiley H, Zebisch M (2018) Climate risk assessment for ecosystem-based adaptation – a guidebook for planners and practitioners. Bonn. https://www.adaptationcommunity.net/wp-content/uploads/2018/06/giz-eurac-unu-2018-en-guidebook-climate-risk-asessment-eba.pdf

International Organization for Standardization (2020) ISO/DIS 14091 Adaptation to climate change — Guidelines on vulnerability, impacts and risk assessment. Geneva

Schneiderbauer S, Zebisch M, Kass S, Pedoth L (2013) 'Assessment of vulnerability to natural hazards and climate change in mountain environments – examples from the Alpse. In: Birkmann J (ed) Measuring vulnerability. United University Press

Zebisch M, Schneiderbauer S, Pedoth L (2014) Regional vulnerability assessment in the Alps. In: Prutsch A, McCallum S, Swart R, Grothmann T, Schauser I (eds) Climate change adaptation manual. Lessons learned from European and other industrialized countries. Tayler & Francis/Routledge

Zebisch M, Schneiderbauer S, Renner K, Below T, Brossmann M, Ederer W, Schwan S (2017) Risk Supplement to the Vulnerability Sourcebook. Guidance on how to apply the Vulnerability Sourcebook's approach with the new IPCC AR5 concept of climate risk'. Bonn. https://www.adaptationcommunity.net/wp-content/uploads/2017/10/GIZ-2017_Risk-Supplement-to-the-Vulnerability-Sourcebook.pdf

Zebisch M, Schneiderbauer S, Fritzsche K, Bubeck P, Kienberger S, Kahlenborn W, Schwan S, Below T (2021) 'The vulnerability sourcebook and climate impact chains-a standardised framework for a climate vulnerability and risk assessment Vulnerability sourcebook and climate impact c, ahead-of-p(ahead-of-print). https://doi.org/10.1108/IJCCSM-07-2019-0042

Chapter 26
Operationalizing Climate Proofing in Decision/Policy Making

Carlo Giupponi

Abstract The purpose of this work is to present an operational approach to include consideration of global change drivers (climatic, economic, social, etc.) in support to the design of local policies or investment plans. In both cases decision/policy makers typically have sets of plausible solutions and decisions to be taken in terms of choices among sets of plausible solutions with the best knowledge about the future dynamics of endogenous and exogenous system variables. The ambition is to identify the preferable solution(s) (in terms of technical performances, acceptance by stakeholders, cost–benefit ratio, etc.) in a medium term perspective, (e.g., 10–40 years), with current knowledge about the problem and under the effect of important sources of uncertainty (both aleatory and epistemic). Common to most decision contexts in a medium term perspective typical of both investment decisions and adaptation policies is the prevalence of economic signals in the shorter term and of climatic signals in the longer term. Models play a fundamental role in both cases, but they rarely cover the whole set of variables needed for decision making and the outcomes usually require integration of qualitative expert knowledge or simply subjective judgements. Multi-criteria analysis coupled with uncertainty analysis can contribute with methodologically sound and operational solutions. This paper elaborates on a series of recent cases with the ambition to extract common elements for a general methodological framework.

Keywords Decision support · Uncertainty · Modelling · Regional scenarios · expert-based assessment

---

C. Giupponi (✉)
Department of Economics, Ca' Foscari University of Venice, and Venice International University, Venice, Italy
e-mail: cgiupponi@unive.it

C. Kondrup et al. (eds.), *Climate Adaptation Modelling*, Springer Climate,
https://doi.org/10.1007/978-3-030-86211-4_26

Introduction

It has become clear that climate changes resulting from the combination of anthropogenic sources (greenhouse gases) and natural dynamics are already affecting social and ecological systems, to which adequate responses must be identified and implemented. The European Union approached the problem through the EU Adaptation Strategy (EU 2013a), with a series of documents and instruments, including the "Guidelines on developing adaptation strategies" (EU 2013b), identifying six main steps of an adaptation process. The three central steps are strongly related to modelling activities (see Fig. 26.1); for example, the assessment of risks and vulnerabilities requires the support of climate change modelling, but also, very importantly, it is only trough integrated modelling that the expected performances of alternative adaptation options can be assessed and thus final decisions about strategies can be taken.

Altered frequencies and magnitude of climate related phenomena (e.g., droughts, storms, floods) affect socio-ecosystems and decision makers have become aware of the importance to include climate risks in medium to long term decisions, both in the policy sector in general and in the financial and economic activities in particular. However, signs of climate change effects always appear in combination with other signals, particularly those deriving from the evolution of markets and policies on different scales (see, e.g., Arnell et al. 2011).

Therefore, entrepreneurs and public decision makers have to define effective development strategies, necessarily taking into account the combined effects of all

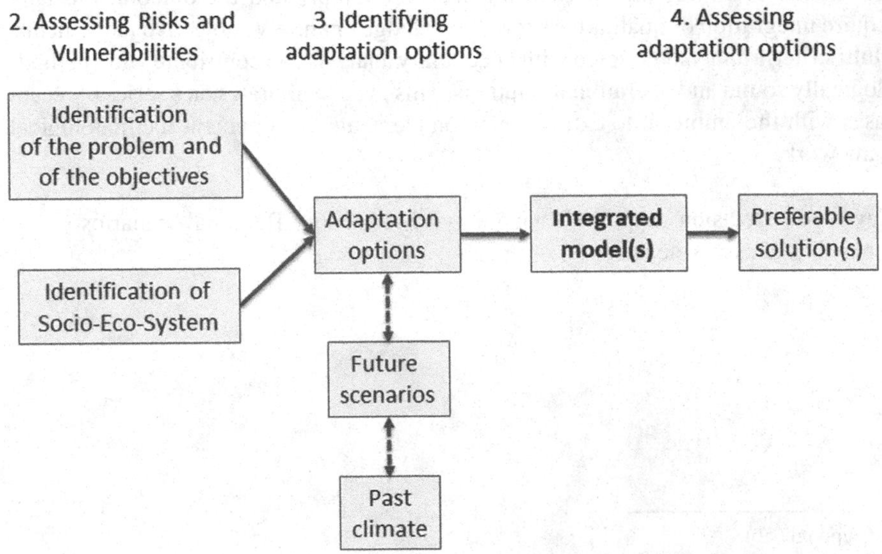

Fig. 26.1 Block diagram showing the contribution of integrated modelling to the identification of adaptation options, following three steps of the EU Guidelines on developing adaptation strategies

the drivers and their dynamics and interrelationships. What they often ask climate change experts is to provide solutions for climate proofing of plans and projects, i.e., to operationalize the available knowledge deriving from climate change integrated modelling and bring it to the decision making process to improve it. In practice, this means to include sources of uncertainty related to the future trajectories of socio-economic development and climate change.

The high level of uncertainty about future evolutions of multiple drivers makes the assessment of risk and resilience a challenging task for analysts and decision-makers. The level of unpredictability of these drivers is known to be deep, since changes have yet to be experienced and knowledge is limited both in terms of modelling (from conceptual to mathematical ones) and in the quantification of these uncertainties. In existing literature, such premises are generally referred to as a situation of deep uncertainty (Lempert and Collins 2007; Lempert and Kalra 2013). Therefore, the need emerges to assist decision-makers not by providing them with an optimal solution based on past trends or a few plausible scenarios, but rather with an analysis able to provide ranges of possible future outcomes under wide sets of plausible scenarios in order to identify robust solutions. As compared to optimal ones, robust solutions are those that show relatively limited cases of failure in a high number of possible future conditions.

A decision-making approach based on the identification of robust solutions (Robust Decision-Making or RDM) has been implemented in several different contexts, such as agriculture, resource management and strategic infrastructure. Public and private investments in maintenance, in protection and in the development of critical infrastructures, i.e., electric power plants, telecommunication and transportation networks, are essential to the functioning of society as a whole. Similarly, effective adaptation policies typically require substantial investment of financial resources, impose chances to consolidated behaviour, and may introduce distributive effects so that in both cases one could say that they should be considered "too important to fail".

A series of recent decision support experiences (private infrastructural and industrial developments, regional policies, etc.) allowed us to extract common needs and solutions and to propose here a methodological framework that could be used to integrate modelling efforts into operational decision making for climate proofing (see, for example, Bernhofer et al. 2019).

The purpose of this work is to present an operational approach to include consideration of global change drivers (climatic, economic, social, etc.) in support to the design of local policies or investment plans. In both cases decision/policy makers typically have sets of plausible solutions and decisions to be taken in terms of choices among sets of plausible solutions with the best knowledge about the future dynamics of endogenous and exogenous system variables. The ambition is to identify the preferable solution(s) (in terms of technical performances, acceptance by stakeholders, cost–benefit ratio, etc.) in a medium term perspective (e.g., 10–40 years), with current knowledge about the problem and under the effect of important sources of uncertainty (both aleatory and epistemic). In both cases decision/policy makes typically have sets of plausible solutions and decisions to be taken in terms of choices. Common to most decision contexts in a medium term perspective typical of both investment decisions and adaptation policies is the prevalence of economic signals in the shorter term and of climatic signals in the longer term. Models play a fundamental role in both cases, but they rarely cover the whole set of variables needed for decision making and the outcomes usually require integration of qualitative expert knowledge or simply subjective judgements. Therefore, both sources of uncertainty should be integrated in the process.

Methods

The proposed approach is aimed in general at analysing alternative options (plans, policies, projects) affected by climate risks or related to climate change adaptation. Such investments are usually characterized by considerable initial and maintenance costs and must therefore be carefully chosen by assessing potential benefits, trade-offs and interactions with the existing settings.

Six main steps are foreseen—developed upon the framework of the EU Guidelines shown in Fig. 26.1—with possible iterations and are depicted in the block diagram of Fig. 26.2.

The first step consist in the identification of the objectives of the actions to be implemented and of the socio-ecosystem (SES) involved, in order to, e.g., identify the boundaries of the system, exogenous and endogenous variable, and the main interacting elements.

In the second step, stakeholders are involved to develop a shared conceptual model of the SES and the main cause-effect relationships between its social, economic and environmental elements, to define the needs for simulation models and other data processing tools, such as spatial analysis ones, together with the required inputs in terms of information to be acquired.

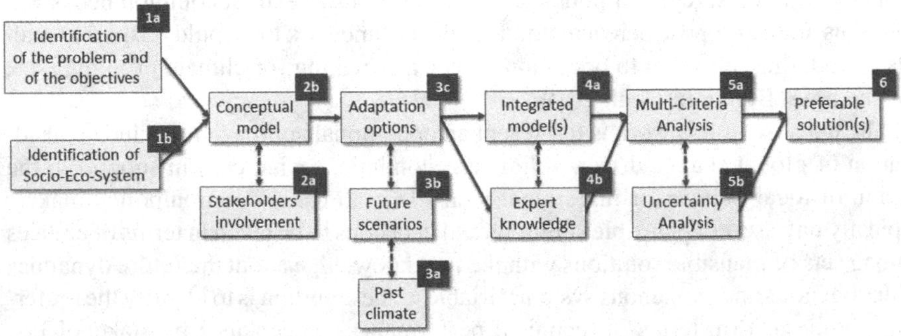

Fig. 26.2 Block diagram for the proposed approach integrating modelling, expert knowledge, scenario analysis, multi-criteria analysis and uncertainty analysis for climate change adaptation

The definition of the conceptual model allows—in a third step—to identify a set of plausible solutions to the given problem and in the specific SES in question. In parallel to that, scenarios to explore how the future may unfold are defined. Exploratory scenarios typically describe plausible trajectories of variables related to different aspects of the future and are used to analyse the possible consequences of predefined assumptions on the evolution of the most important driving forces. In climate change applications, scenarios try to represent the main driving forces deemed relevant so as to bring the correct information into the selection of policies in response to climate changes (Swart et al. 2004; van Vuuren et al. 2012). Scenario development should be done by referring to the most important exogenous variables identified in the first step of the approach, as distinct future situations (e.g., 5 SSP's or N combination of SSP's and RCP's) or, better, as sets of intervals of possible manifestations of the variables. Obviously, the goal in scenario analysis is not to predict the future but to gain a better understanding of possible future alternatives to be able to assess how robust the different decisions or options can be within a wide range of plausible futures. The development of participatory scenarios is increasingly used to stimulate local actors to consider changes that had not been contemplated previously, create integrated images of a future that must be considered in continuous evolution and ensure that multiple skills and subjective interpretations are taken into account, hence strengthening their legitimacy and relevance.

In the following—forth—step, models are utilised, together with expert knowledge. Global models have low resolution as they describe the processes on a continental or a regional scale. In the transition from the global to the local analysis phase (downscaling), many quantitative methods based on mathematical models have been proposed (for an example of integrated models, see Popp et al. 2017). However, the international literature (Lempert et al. 2004; Swart et al. 2004; Alcamo 2008; Van Vuuren et al. 2012) recognises the limits of purely quantitative tools and therefore generally opts for approaches that integrate quantitative analyses with those that use expert judgement (for recent examples, see Palazzo et al. 2017; Kebede et al. 2018). Participatory qualitative scenarios are used to create new ideas and strategies, clarify the options and identify future problems and opportunities, thus incorporating more points of view (Maier et al. 2016).

In the fifth step a classical multi-criteria analysis (MCA) is combined with uncertainty analysis (UA), with which the multiple dimensions of the problem are assessed through decision-relevant indicators (e.g., investment costs, resilience enhancement, environmental impact). In order to offer involved actors methodological solutions with interfaces that could be understood by all, a DSS software was used in our experiences (mDSS; Giupponi 2014). The traditional deterministic MCA is here evolved into a multi-scenario sensitivity analysis by introducing the consideration of various sources of uncertainty (e.g., scenario variables, subjectivity, risk attitude of the decision makers). Numerous MCA matrices are generated to approximate the performance of each alternative and create a range of possible outlooks. The deficiencies of each option under different sets of scenarios are determined; and, based on the above assessment, the robustness of each solution is defined, following the methodology of Rosenhead (1980a, b). In order to provide an effective interface for decision makers, data mining techniques are applied to the multitude of results obtained. The CART (Classification and Regression Trees) algorithm here allows for an identification of critical score values able to overturn the final ranking (step 6) to be shared with the involved actors.

The combination of the various components of the proposed methodology results in a comprehensive and intuitive decision support system that helps the decision-maker to mitigate the impacts of uncertainty and to increase the system's adaptation and resilience to scenario changes, by identifying the most robust solution, which is to be interpreted as the option that ranks best under most of the simulated alternative scenarios.

Applications

Variants of the approach presented above have been applied in various contexts. As applications in support to planning, the Outlook 2030 project and the CORASVE projects financed by the Veneto Region administration can be mentioned. In the first case, sets of measures in consideration for the future Rural Development Plan were analysed through a sequence of expert workshops, vis a vis the alternative scenarios deriving from downscaling of IPCC SSP's at regional level. Similarly, in the CORASVE Project the approach was applied in support of the analysis of measures proposed for a general conference on agriculture, still in support to the RDP. Among projects in support of private investments and planning, applications were in the field of climate proofing of hydraulic safeguarding of infrastructures, decarbonisation and renewable energies, and other economic activities, such as tourism and electric power distribution, all with the involvement of the main local stakeholders (local administration, SMEs, big farm, port authority) in the assessment of strategies under the effect of future climatic scenarios of normative and market evolutions.

Conclusions

Decision making for climate change adaptation is affected by deep uncertainty, i.e., by the lack of agreement about how the future will look like, about probability distributions and about parameter values. This has serious consequences for modelling exercises that can only partially dealt with by running model ensemble simulations. Moreover, climate change drivers must be considered jointly with others and in particular with socio-economic ones.

Outputs of modelling exercises cannot be immediately used by decision/policy makers; instead, they have to be integrated with other sources of knowledge (local, collective, subjective, qualitative,…) and jointly implemented within an integrated platform for decision support.

A combination of qualitative and quantitative modelling, with multi-criteria analysis and data mining techniques, can significantly improve the potentials of modelling techniques alone.

Hence, instead of following the traditional path of assessing discrete and deterministic, or in some cases probabilistic, values and searching for optimal solutions, decision-making for climate change adaptation requires lying out all the conditions under which plausible solutions may emerge and search for more robust options. Decision-makers are thus informed about how far unknown future events led by various related or isolated factors may influence their ability to adapt and cope with the negative consequences and the positive opportunities that may arise.

References

Alcamo J (2008) The SAS approach: combining qualitative and quantitative knowledge in environmental scenarios. In Joseph A (ed.) Environmental futures. The practice of environmental scenario analysis. Elsevier Oxford, UK, pp 123–150

Arnell NW, van Vuuren DP, Isaac M (2011) The implications of climate policy for the impacts of climate change on global water resources. Global Environ Chang 21:592–603

Bernhofer J, Giupponi C, Mojtahed V (2019) A decision-making model for critical infrastructures in conditions of deep uncertainty. In: Cecconi F, Campennì M (eds) Information and communication technologies (ICT) in economic modeling. Springer International Publishing, Cham, pp 139–161

EU (2013a) An EU strategy on adaptation to climate change. COM(2013) 216 final

EU (2013b) Guidelines on developing adaptation strategies. SWD(2013) 134 final

Giupponi C (2014) Decision support for mainstreaming climate change adaptation in water resources management. Water Resour Manag 28(13):4795–4808

Kebede AS, Nicholls RJ, Allan A, Arto I, Cazcarro I, Fernandes JA, Chris T, Hill CT, Hutton CW, Kay S, Lázár AN, Macadam I, Palmer M, Suckall N, Tompkins EL, Whitehead PW (2018) Applying the global RCP–SSP–SPA scenario framework at sub-national scale: a multi-scale and participatory scenario approach. Sci Total Environ 635:659–672

Lempert R, Nakicenovic N, Sarewitz D, Schlesinger M (2004) Characterizing climate-change uncertainties for decision-makers. An Editorial Essay Climatic Chang 65:1–9

Lempert RJ, Collins MT (2007) Managing the risk of uncertain threshold responses: comparison of robust, optimum, and precautionary approaches. Risk Anal 27(4):1009–1026

Lempert RJ, Kalra N (2013) Managing climate risks in developing countries with Robust decision making, Washington DC

Maier H, Guillaume J, van Delden H, Riddell GA, Haasnoot M, Kwakkel JH (2016) An uncertain future, deep uncertainty, scenarios, robustness and adaptation: how do they fit together? Environ Modell Softw 81:154–164

Palazzo A, Vervoort JM, Mason-D'Croz D, Rutting L, Havlík P, Islam S, Bayala J, Valin H, Kadi AHK, Thornton P, Zougmore R (2017) Linking regional stakeholder scenarios and shared socioeconomic pathways: quantified West African food and climate futures in a global context. Global Environ Chang 45:227–242

Popp A, Calvin K, Fujimori S, Havlik P, Humpenöder F, Stehfest E, Bodirsky BL, Dietrich JP, Doelmann JC, Gusti M, Hasegawa T, Kyle P, Obersteiner M, Tabeau A, Takahashi K, Valin H, Waldhoff S, Weindl I, Wise M, Kriegler E, Lotze-Campen H, Fricko O, Riahi K, van Vuuren DP (2017) Land-use futures in the shared socio-economic pathways. Global Environ Chang 42:331–345. https://doi.org/10.1016/j.gloenvcha.2016.10.002

Rosenhead J (1980a) Planning under uncertainty: II. A methodology for robustness analysis. J Oper Res Soc 31(4):331–341. http://www.jstor.org/stable/2581626

Rosenhead J (1980b) Planning under uncertainty: 1. The inflexibility of methodologies. J Oper Res Soc 31(3):209. https://www.jstor.org/stable/2581077

Swart RJ, Raskin P, Robinson J (2004) The problem of the future: Sustainability science and scenario analysis. Global Environ Chang 14(2):137–146

van Vuuren DP, Kok MTJ, Girod B, Lucas PL, de Vries B (2012) Scenarios in global environmental assessments: Key characteristics and lessons for future use. Global Environ Chang 22(4):884–895

Chapter 27
Adaptation Modelling: A JPI Climate Perspective

Alexandre F. Fernandes and Frank McGovern

Abstract The Joint Programming Initiative "Connecting Climate Knowledge for Europe" (JPI Climate) is a pan-European intergovernmental initiative gathering European countries to jointly coordinate climate research and fund new transnational research initiatives that provide useful climate knowledge and services for post-COP21 Climate Action. The main objective of JPI Climate is to bring together existing and developing new excellent scientific knowledge that is needed to assist practitioners to adequately transform society towards climate resilience and consequently providing integrated climate knowledge and decision support services for societal innovation. To date, JPI Climate has mobilised more than 100 million EUR in research investments and has provided access to knowledge and expertise across Europe and beyond. Some of the key projects from JPI Climate include "European Research Area for Climate Services" (ERA4CS), designed to boost the development of efficient climate services, "Assessment of Cross(X)-sectoral climate Impacts and pathways for Sustainable transformation" (AXIS), which aims to promote cross-boundary, cross-community research with the overall goal to improve coherence, integration and robustness of climate impact research and connect it to societal needs, and "Enabling Societal Transformation in the Face of Climate Change" (SOLSTICE), bringing together the Social Sciences and Humanities communities to enable and accelerate positive transformation in the face of climate change. The current development of JPI Climate Knowledge Hubs and the potential establishment of a European Facility for Climate Change (EFCC) will further establish JPI Climate as a key player in European climate change research and will actively inform and support the implementation of relevant national, European and international climate strategies and policies.

Keywords JPI climate · Climate services · Science-policy interface · Knowledge Hubs · EFCC

A. F. Fernandes (✉)
JPI Climate Central Secretariat, Belgian Science Policy Office (BELSPO), Brussels, Belgium
e-mail: alexandre.fernandes@jpi-climate.belspo.be

F. McGovern
Environmental Protection Agency, Chair of the JPI Climate Governing Board, Wexford, Ireland
e-mail: f.mcgovern@epa.ie

C. Kondrup et al. (eds.), *Climate Adaptation Modelling*, Springer Climate,
https://doi.org/10.1007/978-3-030-86211-4_27

Introduction

The Joint Programming Initiative "Connecting Climate Knowledge for Europe" (JPI Climate) is an initiative of EU Member States and Associated Countries, in cooperation with the European Commission (EC). JPI Climate, comprised of representatives of ministries and organisations for research funding, aims through its programme of activities to connect research, performers and funders across Europe to promote the creation of new knowledge in the natural and anthropogenic climate change domain that is fundamental and relevant for decision support.

The vision of JPI Climate is to actively inform and enable the transition to a low emission, climate resilient economy, society and environment that is aligned with Europe's long-term climate policy objectives. JPI Climate shall therefore develop and coordinate a pan-European research programming platform to provide useful climate knowledge and services for European and national climate strategies and plans and contributions to the United Nations Framework Convention on Climate Change (UNFCCC) and the United Nations Sustainable Development Goals (SDGs).

JPI Climate's mission is to align and inform strategies, instruments, resources and actors at national and European levels by connecting the various research communities with research funders and performing organisations, within and across European countries, and beyond Europe.

JPI Climate's vision and mission are framed in its Strategic Research and Innovation Agenda (SRIA) (JPI Climate, 2016), which also sets out three overarching challenges and one strategic mechanism that together are intended to develop and support excellent, innovative, relevant and informative climate research.

Since its establishment, JPI Climate has been at the centre of pan-European investments in climate change research and in harnessing their outcomes to inform effective responses by policymakers and practitioners. Ultimately, these efforts are designed to underpin the European efforts in tackling climate change.

Key Projects

To date, JPI Climate has mobilised more than 100 million EUR in research investments and has provided access to knowledge and expertise across Europe and beyond. This has been possible with the support of its member countries and the EC, and it has been done in partnership with other JPIs (such as FACCE-JPI, JPI Oceans, JPI Urban Europe) and, at the global level, with the Belmont Forum. A list of projects that have been (or are currently being) funded by JPI Climate is available in the JPI Climate e-magazine (JPI Climate, 2019).

In 2013, JPI Climate published its first joint call for transnational collaborative research projects aiming to provide support for top-quality research projects on two topics: Societal Transformation in the Face of Climate Change and Russian Arctic and Boreal Systems.

Projects awarded under the first topic aimed to inform and support societal transformations in the face of climate change and in line with sustainable development in Europe and globally. For the second topic, the awarded projects aimed to improve the fundamental understanding of key biological and physical drivers and feedbacks in the Russian Arctic/Boreal system (tundra-taiga-coastal region) to enable better representation of these processes in climate models.

In 2016, JPI Climate launched "European Research Area for Climate Services" (ERA4CS), its flagship project on climate services (in collaboration with the EC).

The overall objective of ERA4CS is to enhance user adoption of and satisfaction with climate services (including adaptation services). It implies the development of tools, methods and standards to produce reliable information (projections with impact and vulnerability assessments) for various user needs to support smart stakeholder decisions and investment projects, encompassing public, private and community sectors.

At the same time, ERA4CS aims to improve the scientific expertise on climate change risks and adaptation options and to connect that knowledge with decision-making, e.g., by developing and assessing climate adaptation strategies and pathways at different scales (regions, cities, catchments, vulnerable sectors, etc.).

It focuses on the development of a "climate information translation" layer, including climate information production for climate services, as well as researching and advancing climate services as such.

The JPI Climate project "Assessment of Cross(X)-sectoral climate Impacts and pathways for Sustainable transformation" (AXIS) (in collaboration with the EC) aims to promote cross-boundary, cross-community research with the overall goal to improve coherence, integration and robustness of climate impact research and connect it to societal needs. To this effect, AXIS aims to overcome boundaries between science communities through inter- or trans-disciplinary research projects.

AXIS is a successor of ERA4CS. Both are part of the efforts of JPI Climate to contribute to the implementation of the European Roadmap for Climate Services.

The projects funded by JPI Climate range from advancing the understanding of fundamental climate science to enabling the societal transformations that are required in the face of climate change. In this context, the project "Enabling Societal Transformation in the Face of Climate Change" (SOLSTICE), recently launched, aims to bring together the Social Sciences and Humanities communities to enable and accelerate positive transformation in the face of climate change, by engaging with societal actors through innovative approaches.

Ongoing and Future Developments

Since its foundation, JPI Climate has established itself as a key player in the development of pan-European research on climate change through joint funding for research projects of shared interest.

A key objective of all research investments is to ensure a return and valorisation of the outcomes. This has a particular urgency for investments in climate change research and innovation. The returns for investments can include:

New knowledge and progress in understanding the causes of climate change;

Uptake or use of information and findings in decision-making and policy development;

Development and use of solutions by practitioners and stakeholders.

At the global level, the Intergovernmental Panel on Climate Change (IPCC) has provided large scale assessments of progress in understanding climate change and options to address its causes and consequences. These are communicated to policy-makers every 6 to 7 years. Although some countries carry out similar assessments at the national level, there is currently no similar assessment and communication process at the European level, which represents a key operational and strategic gap.

To address this gap, JPI Climate has decided to establish and progress the development of two pilot Knowledge Hubs (KH): one on climate neutrality; and another one on sea level rise.

These will address issues that are central to the science-policy interface and to practical responses in addressing climate change, from local to global scales. Information on these issues is also increasingly and urgently needed to provide effective responses to climate change.

The objectives of the KH on climate neutrality are to provide regular authoritative analysis of the scientific understanding of climate neutrality, to progress on pathways towards achievement of such goal, to provide updates on the potential contribution of technologies and innovative socio-economic and transdisciplinary approaches to advance this process, and to identify knowledge gaps and research needs. The assessments will be global in nature, but focused on Europe and the specific issues and challenges that exist in Europe, including those that arise from its diverse development pathways, geographical and climatic diversity, and societal complexities.

The long-term ambition of the KH on sea level rise is to provide periodic assessments of knowledge on sea level rise drivers, impacts and policy options for each of the major ocean basins around Europe. It will complement existing global and European assessments by providing additional geographical and contextual detail, tailored to regional, national and European policy development and implementation.

These KH will be part of an anticipated future European Facility for Climate Change (EFCC), an innovative organised structure and process supported by governments to provide authoritative knowledge that is designed to address policy and practitioner needs and inform decision-making in an open and dynamic manner. The goal of the EFCC is to support strategic joint flagship projects, clusters of projects or programmes that are able to fill critical knowledge gaps in climate change research or spark developments of networks, knowledge communities or science fields particularly relevant for the implementation of the Paris Agreement and the SDGs.

Conclusions and Recommendations

By aligning the collective investments from European countries in climate change research and innovation and ensuring the uptake of the outcomes in decision-making and policy development, JPI Climate plays a crucial role in informing and supporting the implementation of relevant national, European and international climate strategies and policies.

In this context, JPI Climate is actively engaged in the discussions informing the development and implementation of relevant European climate policy processes, including Horizon Europe and its Missions (Adaptation to climate change, including societal transformation; Climate-neutral and smart cities; Soil health and food), the European Green Deal, the European Climate Law, and the new EU strategy on adaptation to climate change.

JPI Climate is also leading the implementation of the next European Climate Change Adaptation (ECCA 2021) conference, in collaboration with the EC, which will provide a key platform to bring scientists, policy-makers and practitioners together to advance the Mission on adaptation to climate change and showcase JPI Climate's flagship work on climate services. Realising the potential of the ECCA process can be an important step in the development of a KH on solutions and services for climate resilience.

Through its current and future activities, JPI Climate remains committed to provide the knowledge base needed to inform climate policy and decision-making in Europe and beyond.

References

JPI Climate (2016) Strategic Research and Innovation Agenda. Joint Programming Initiative Connecting Climate Knowledge for Europe, Brussels, Belgium. http://www.jpi-climate.eu/gfx_content/documents/JPI%20CLIMATE_SRIA_LR.pdf. Accessed 17 July 2020

JPI Climate (2019) JPI climate e-magazine. http://www.jpi-climate.eu/media/default.aspx/emma/org/10900194/JPI_Climate_e-magazine_17Feb2020.pdf. Accessed 17 July 2020

Correction to: The Roles of Climate Risk Dynamics and Adaptation Limits in Adaptation Assessment

Laurens M. Bouwer

Correction to:
Chapter 24 in: C. Kondrup et al. (eds.), *Climate Adaptation Modelling*, Springer Climate,
https://doi.org/10.1007/978-3-030-86211-4_24

The original version of the book was published with an incorrect figure, Figure 24.1 has been updated. The chapter and book have been updated.

The updated version of this chapter can be found at
https://doi.org/10.1007/978-3-030-86211-4_24

C. Kondrup et al. (eds.), *Climate Adaptation Modelling*, Springer Climate,
https://doi.org/10.1007/978-3-030-86211-4_28

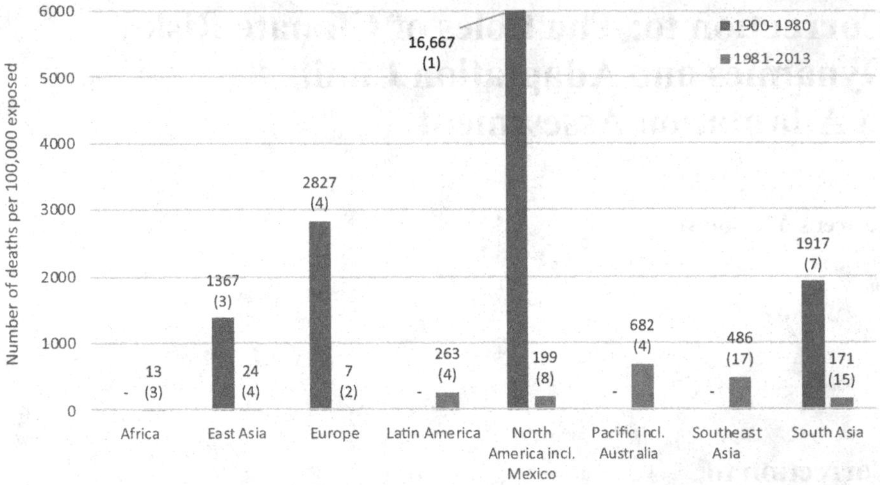

Fig. 24.1 Changes in average event mortality (number of deaths per 100,000 exposed population) for storm surge floods for different world regions between 1900 and 2013. The number of included events is given in brackets. Data from Bouwer and Jonkman (2018)

Printed in the United States
by Baker & Taylor Publisher Services